Why Are Orangutans Orange?

Science questions in pictures –
with fascinating answers

Why Are Orangutans
Orange?

Science questions in pictures –
with fascinating answers

edited by Mick O'Hare

PEGASUS BOOKS
NEW YORK LONDON

Why Are Orangutans Orange?

Pegasus Books LLC
80 Broad Street, 5th Floor
New York, NY 10004

Copyright © 2012

ISBN: 978-1-60598-389-9

10 9 8 7 8 6 5 4 3 2 1

Printed in the United States of America
Distributed by W. W. Norton & Company, Inc.
www.pegasusbooks.us

Contents

Introduction

OK, we admit it, they're cute. That's why they are on the front of the book. But when asked, we had no idea why orangutans were a strange orange colour – one that didn't even seem to match their environment. It was a long time before we received any response to the question too, suggesting that even the experts were a little unsure. But now we think we know – turn to page 30 to find out.

You'll also discover among these pages why tigers have stripes rather than spots, why blue-footed boobies have, erm, blue feet and whether kittiwakes can fly upside down. And it's not all about animals. We have the lowdown on any number of clouds, strange bubbles and weird ice... and all in glorious colour.

Readers of our earlier books such as *Why Don't Penguins' Feet Freeze?* and *Does Anything Eat Wasps?* will notice a difference in this latest collection of questions and answers from *New Scientist*'s Last Word column – photographs: and lots of them.

Most haven't been supplied by professionals. In fact, nearly all have been taken by readers of The Last Word column in *New Scientist* magazine and on its website. Some of the photographs are extraordinary, many are unique and some are a bit fuzzy. But we can handle that because they tell the visual story of some extraordinary phenomena, taken on the run by members of the public.

And that is the essence of this book: a celebration of the wonder of our world that any inquisitive person lucky enough to be in the right place at the right time can witness and record – if we are prepared to keep our eyes open. Specialists in their fields have spent years waiting to capture these moments but they have been beaten to it by readers of *New Scientist* and its books.

All of which means we can now tell you the story of why flies sometimes explode, what you should do when your hair stands on end (and why it's very, very important to do it quickly) and why Mount Fuji sometimes appears to be wearing a hat.

If you have any similar images that you have captured somewhere in the world – from your back garden to coldest Antarctica – and have always wondered what on earth they show, The Last Word can help. Every week hundreds of questions pour into our offices, some with photographs, others without. You can add yours to the list, or help us answer the ones we are still puzzling over. Visit www.newscientist.com/lastword to ask a question, or help us answer one. And buy the magazine to check out our weekly page. You could even appear in the next book (or at least your photograph could).

Mick O'Hare

1 All creatures great and peculiar

? Happy feet

The blue-footed booby is an extraordinary-looking bird. It has fairly dull plumage but strikingly coloured blue legs and feet. What could be the evolutionary benefit of such a conspicuous feature? Both sexes have blue feet so they don't seem to be for impressing potential mates.

Sam Moore
London, UK

Although not obvious at first sight, during courtship blue-footed boobies (*Sula nebouxii*) have different-coloured feet depending on their sex: male feet are brighter and more of a greenish-blue, while the females have duller feet that are bluish.

The birds exhibit their feet to prospective partners in a series of courtship displays. These include a kind of ritualised strutting around that allows them to show off their feet, plus stylised or 'salute' landings which serve the same purpose.

I am a member of a research group that studies the sexual behaviour of the blue-footed booby. In one experiment, we altered the colour of the courting males' feet and recorded the females' response. Females paired to males with duller feet were less enthusiastic about courtship and less likely to copulate compared with females paired to males with normal, brightly coloured feet. Similarly, when we altered the females' feet to a duller blue, males became less interested in courting them. Birds in poor health often have dull blue feet.

What's more, females whose mates had dull blue feet produced smaller eggs, and their chicks had a poorer immune response compared with normal females. This may sound surprising, but it is in accordance with theoretical expectations.

All this suggests that males are probably under strong selection pressure to maintain greenish-blue feet during courtship. This will ensure not only that they copulate successfully but also that their mates will lay big, healthy eggs. Overall, our results suggest that foot colour is a trait maintained by mutual male and female preferences.

Roxana Torres
Institute of Ecology
National Autonomous University of Mexico

Both male and female *Sula nebouxii* have blue feet, but it is the male that presents his feet prominently in courtship. This, in effect, is a way of saying that he is of the same species as the female.

I cannot offer any specific reason why the blue-footed booby has blue feet, but I would point out that foot colour does seem to be significant in the genus – there is an equally striking red-footed booby, *Sula sula*. This suggests that as members of the genus evolved, they adapted to different ecological niches which, in turn, meant that there was an advantage in the birds splitting into different 'tribes' that could only mate with their own kind.

This is an example of what is called sympatric evolution, where one species evolves into two within a shared territory. In contrast, allopatric evolution occurs because populations become isolated from each other. For sympatric evolution to succeed, it is essential that some sort of difference between the species arises so that a bird can distinguish between a bird of a related species and one of its own kind.

Guy Cox
Associate Professor
Australian Centre for Microscopy & Microanalysis
University of Sydney, Australia

❓ Ducking the issue

I have never seen a duck stand as erect as the one shown in the centre of this photo, which I took at Rowsley, Derbyshire. Does anyone know if there is an explanation for this posture or is it just an unexpectedly tall duck?

Vince Sellars
Sheffield, UK

Most of the birds in the background of the photograph are male and female mallards (*Anas platyrhynchos*) from which almost all domestic ducks originate and hence commonly and freely interbreed.

But the upright drake is a cross-breed – note the less-clearly defined markings compared with the other drakes. He is half mallard and his other parent was an Indian runner. This is a common breed that is raised for its egg-laying performance and is characterised by its distinctive vertical stance and slender frame, which results in a comical gait. Standard domestic ducks similar to the others in the photograph, which are bred for their meat, retain a more normal horizontal carriage.

Interestingly, the slender upright stance seen in this duck is quite dominant genetically, and interbreeding between Indian runners and other ducks typically results in skinny, upright offspring. Indian runners come in a wide variety of colours, with white and brown being the most common.

Giles Osborne
Mitcham, Surrey, UK

The erect duck is a hybrid of a mallard duck and a domestic Indian runner duck. Indian runners and the crested version, Bali ducks, came from Indonesia – not India – and were brought to Europe by Dutch traders. They were once known as penguin ducks because of their erect stance.

Gail Harland
Coddenham Green, Suffolk, UK

For those who would like to explore the parentage and history of this bird further, check out the Indian Runner Duck Association at www. runnerduck.net. Thanks to Jo Horsley of Llanwrda, Carmarthenshire, UK, and others for pointing this out – Ed.

? Off-centre

With the exception of the sperm whale's off-centre blow-hole and some crabs' single large claw, all complex organisms I can think of are effectively symmetrical along one plane of their body. What is the least symmetrical organism?

Max Maguire
By email; no postal address supplied

Flatfish received the largest vote, but there are plenty of other strange candidates out there – Ed.

The least symmetrical organism is the halibut, which has both eyes on the same side of its head.

Donald Windsor
Norwich, New York, US

There are different symmetries in nature. We tend to assume bilateral symmetry is normal because that is what we and most of the organisms we notice (vertebrates and arthropods) display. But bilateral symmetry is the exception rather than the rule; many creatures exhibit radial or even spherical symmetry. Some alter their symmetry over time – for example, a starfish will start out as a bilaterally symmetrical larva and become radially symmetrical as it matures. Humans, a few days after conception, are basically a spherically symmetrical organism called a morula.

Many organisms do not have any clear geometrical symmetry but demonstrate some kind of fractal symmetry, where their structures look similar at a variety of scales. Many plants and fungi are a bit of both: think of the leaves and the apples on an apple tree.

Humans are not quite bilaterally symmetrical: our liver is on the right side, our spleen on the left, while our right lung has three lobes and the left two. We even slip into fractal symmetry when it suits the purpose: take a close look at the capillaries which transport blood to the tissues. We are not even superficially symmetrical. Next time you get out of the bath ask yourself, 'Do they both hang the same?' This works for either sex.

We are all changed and shaped by both our genes and our environment. To put it another way, we all conform to a pattern while being eccentric. Heck, that's life.

David Hopkins
Smethwick, West Midlands, UK

Members of the genus *Histioteuthis*, squid that live down to depths of 1000 metres, are unique in the animal kingdom as their left eye is two to three times the size of the right. The reasons for this trait, which gives rise to its common name of the cock-eyed squid, are unclear. There is also a corresponding asymmetry in the optic lobes of the squid's brain. The specimen pictured below was filmed on board ship after being caught off the coast of California.

Ron Douglas
Department of Optometry & Visual Science
City University, London, UK

One suggestion is that the depth at which cock-eyed squid live is about as far down as sunlight can penetrate. The squid trains one eye on the illuminated water above while the other looks down into blackness – Ed.

Asymmetry is commonest among organisms that have little need of well-defined structures in their bodies. Some algae, fungi and sponges never developed much symmetry, while parasites can abandon symmetry when they grow opportunistically to secure food. An example of the latter is *Sacculina*, a barnacle that injects its soft body through a crab's shell and then grows a lump of reproductive tissue plus a tangle of feeding filaments throughout the crab's body. And some members of the group of tiny crustaceans known as copepods form shapeless reproductive sacs within cysts in the flesh of fishes. Such creatures need no symmetry.

Jon Richfield
Somerset West, South Africa

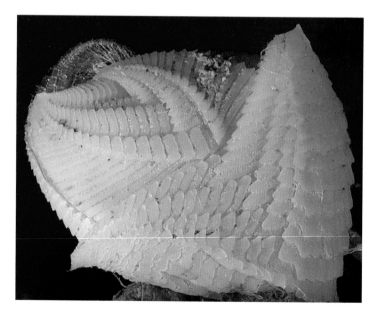

There is a wonderfully quirky group of asymmetric barnacles called verrucomorphs, shown in the photograph opposite. They are either 'right-handed' or 'left-handed' – apparently a random choice – their form being determined by the loss of calcareous plates from either the right or left side of the shell wall, plus a reduction in the number of plates in the shell's lid to two from the usual four. Why they have adopted this asymmetric form is uncertain, particularly as both forms occur together. Their soft tissue, as it happens, retains bilateral symmetry. These and other asymmetric barnacles were first comprehensively described by Charles Darwin in a monograph published in 1854.

John Buckeridge
RMIT University
Melbourne, Australia

? Eggstraordinary

One of my young chickens has just produced an unusually coloured egg (at the right of the photo). The egg on the left is more typical of the breed. I know egg shell colour is variable, even in eggs laid by the same hen on different days, but how did one egg undergo such a sudden and distinct colour change?

Colin Booth
Durham, UK

The answer probably lies in the fact that, until shortly before they are laid, hens' eggs are white. The brown pigmentation associated with breeds such as the Rhode Island red and the maran is a last-minute addition during egg formation and, like a fresh coat of paint, can come off surprisingly easily.

More than 90 per cent of the shell of a hen's egg consists of calcium carbonate crystals bound in a protein matrix. The shell starts to form after the egg has reached the uterus, where it stays for around 20 hours prior to being laid.

During this time, glands secrete the shell around the membranes that hold the yolk and albumen. In brown-egg-laying breeds, the cells lining the shell glands release pigment during the last 3 to 4 hours of shell formation. Most of the pigment is transferred to the cuticle, a waterproof membrane that surrounds the porous eggshell.

Several factors can disturb the cuticle formation process and thus pigmentation, such as ageing, viral infections – including that perennial chicken farmer's nemesis, bronchitis – and drugs such as nicarbazin, which has been widely fed to poultry to combat a disease caused by a type of protozoan. Possibly the most significant factor affecting egg pigmentation is exposure to stress during the formation of the egg.

If a flock of hens is disturbed by a fox during the night, for example, they might well lay paler eggs in the morning. The adrenaline the hens release puts egg-laying on hold and shuts down shell formation. The egg's pigmentation will be affected if the cuticle doesn't form properly.

Even if the pigment is laid down, there is no guarantee that it will last, as Morris Steggerda and Willard F. Hollander found in 1944 while they were studying eggs from a flock of Rhode Island reds in the US. When they cleaned the eggs, the brown pigment occasionally came away; the harder the eggs were rubbed, the more pigment was removed. Only those shells with a glossy sheen retained their colour, suggesting their cuticles had been fully formed, with a protective layer that acted rather like the varnish on an oil painting.

As for the egg photographed by your questioner, the bird was probably disturbed while the cuticle was being formed and so the pigment, inadequately protected, was rubbed off the larger, rounder end of the egg as it was forced out.

The issue may have some significance for public health. The waterproof cuticle is the egg's defence against bacteria. As shell colour is affected by how well the cuticle forms, it also provides a visual test of how free from harmful bacteria an egg may be.

Hadrian Jeffs
Norwich, Norfolk, UK

Before an egg is laid, the hen's shell gland secretes pigment into the fluid bathing the egg's surface. The fluid smears readily, and any disturbance while the egg dries can create marks. Farmers are therefore fussy about the kind of bedding they use in nesting boxes.

Eggs are usually laid big end first. The hen that laid the egg in question may have resorted to using friction to release the partly laid egg from its cloaca, possibly by rubbing the egg against the bedding it was sitting on.

Alternatively, the hen may have paused halfway through laying, perhaps disturbed or exhausted, with the egg half-protruding from its cloaca. The part of the egg still within the cloaca had time to achieve a deep colour before the hen relaxed again, and this accounts for the sharp boundary in coloration seen in the photo. Such a scenario is unusual but it does happen.

Jon Richfield
Somerset West, South Africa

❓ Life on Uluru

Some decades ago I was lucky enough to climb Uluru in Australia's Northern Territory. Recent rain had left pools on top of the rock and, curiously, in many of them there were strange aquatic invertebrates as seen in the photo. They look like ancient trilobites. Why and how are they on top of the famous, massive rock, and what are they? What happens to the creatures when the puddles dry up?

Gavin Chester
Dwellingup, Western Australia

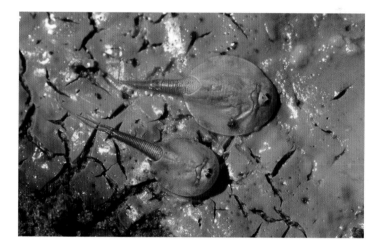

The animals pictured are shield shrimps, *Triops australiensis*. They are crustaceans in the class Branchiopoda – meaning 'gill-legged' – and this term reflects the fact that they use their legs for breathing as well as for movement.

Their external morphology appears to have remained unchanged for 220 million years or more, and one shield shrimp, *Triops cancriformis*, has been claimed by some to be the oldest extant animal species. They occur in bodies of fresh or slightly salty water that periodically dry out, such as ephemeral lakes, farm dams, ditches and even puddles left after rain.

The eggs of these animals have a very strong shell and are resistant to drying out. In some species, a period of desiccation is necessary for the creature's development. The eggs can tolerate freezing and temperatures up to 80 °C, and may remain viable for 25 years. In some species, hatching may take up to a year following exposure to suitable conditions, but in *T. australiensis* it usually takes several weeks at most. Once hatched, development from egg to adult may take only a further few weeks in summer temperatures. The animals have a lifespan of up to three months, and adults reach a length of about 35 millimetres.

The shrimps feed on microscopic organisms, aquatic worms, other shrimp species, frogs' eggs and tadpoles, decomposing vegetation and other detritus, and sometimes even moulting individuals of their own species. The small size and the robustness of the eggs allow them to be carried on the wind for hundreds of kilometres from their pools of origin, and it is probably this mode of transport that would have delivered the eggs to the top of Uluru.

It is also possible that the eggs might have been carried up in mud caked on a visitor's boots. Although in this instance such a method of transport is essentially innocuous, it is nevertheless a salient reminder of the need to ensure that all clothing and equipment is cleaned before moving from one ecosystem to another.

Harko Werkman
Woodbridge, Tasmania, Australia

I bought a packet of desiccated shield shrimp eggs (*Triops australiensis*) on the internet for my boyfriend's 30th birthday. As the name *Triops* suggests, shield shrimps have three eyes: two compound eyes and one 'nauplar eye' – a simple median eye, first appearing in the larval stage. They closely resemble their Triassic ancestors, which existed around 220 million years ago.

Blown around with the dust, eggs eventually settle in crevices and grooves – even on the top of the great rock – where they may remain viable for up to 10 years. I guess that means my boyfriend has an excuse for not hatching them yet.

Kate Hutson
School of Earth and Environmental Sciences
University of Adelaide, Australia

Triops species are found on most continents but are rare in the UK, where the tadpole shrimp (*Triops cancriformis*) is currently known to exist in only two locations, the New Forest in southern England and the Solway Firth in southwest Scotland. Visitors to the Wildfowl and Wetlands Trust Centre at Caerlaverock, Dumfriesshire, can view this species in the visitor centre.

Triops is typical of ephemeral, or temporary, wetlands, and can survive drying to persist for up to 30 years as eggs or cysts. The eggs at Caerlaverock were collected to provide a safety net for the population in Scotland, where they persist in one temporary pond which has been created by cattle trampling around a fence post.

Emma Hutchins and Sally Cordwell
Wildfowl and Wetlands Trust
Slimbridge, Gloucestershire, UK

❓ Bird on a wire

I saw this kittiwake flying upside down in Norway's Svalbard
archipelago – about 79° North – while I was stuck in the sea ice. This
and other kittiwakes were feeding on polar cod (about 13 centimetres
long) that had become uncovered as our ship broke through the ice.
What is it doing and why? How many other birds can do this?

Bill Reed
US

The kittiwake (*Rissa tridactyla*) is not flying upside down at
all. You can tell because the bird's upper wings are visible in
the photo, showing the silvery-grey feathers with the classic
dipped-in-black-ink wing tips. If it were upside down we would
see the underwings, which are white with black tips. The twist
of the bird's head is interesting, though. Clearly the bird has
turned its head a long way to the right, so that it seems initially
to be flying upside down.

Many birds can rotate their heads to this degree or more

– owls and other birds of prey are the best-known examples. In these species, head-turning helps them to detect their prey. Specifically, it allows owls to orientate their ears to obtain the best possible reception when listening for the faint rustling of a rodent moving through vegetation in the dark.

In the kittiwake, however, this doesn't happen. A possible explanation is that the bird is trying to cough up a particularly sharp piece of fish bone or something else it has swallowed. Many birds, including gulls, regurgitate indigestible pellets.

Another explanation is that the bird is shaking off excess salt water from its beak. Most seabirds take in varying amounts of salt water when feeding, which they have to get rid of before it reaches harmful levels in the body. Finally, the bird may simply be twisting as it calls out to other individuals in the same area, or just keeping a lookout for potential predators, such as skuas.

Kevin Elsby
Norwich, Norfolk, UK

If you get a chance to look at the photos in the websites below (especially the first) do so – Ed.

The bird is flying normally and twisting its head around, perhaps to preen itself or to loosen a morsel of fish that it may be eating. The underwing pattern of the kittiwake looks nothing like the upper wing at all, and a cursory inspection of the structure of the flight feathers of this bird reveals a normally aligned gull.

Birds do not generally fly upside down, but they may momentarily invert, such as when wildfowl 'whiffle' to lose height rapidly, spilling air from under their wings (see bit. ly/3wfoqV). Additionally, some birds may roll during mating displays, such as the aptly named roller birds, part of the order Coraciiformes, while others might in play (see bit.ly/9oq8PT).

Simon Woolley
Winchester, Hampshire, UK

❓ All alone

Do polar bears get lonely? I'm not being flippant, just attempting to find out why animals such as humans or penguins are gregarious while others, such as polar bears and eagles, live more solitary lives.

Frank Anders
Amsterdam, The Netherlands

Readers will recognise this question as the title of one of our earlier books, Do Polar Bears Get Lonely? *While we are not going to repeat all the answers here – you'll have to buy the book if you haven't already – we were impressed with this photograph of a polar bear and a husky, taken by Norbert Rosing. They are seemingly getting along just fine. If polar bears do indeed get lonely, every so often, it seems, they are prepared to cavort with their fellow animals in the Arctic – Ed.*

❓ Batty behaviour

About a million bats fly out of the Mulu Cave in Sarawak, Malaysia, every evening to look for food. They fly out in batches and it sometimes takes two hours before they are all in the air. When they leave the cave they form a circle before forming sinusoidal waves that stretch great distances. Why do they fly like this?

Mazlan Othman
Putrajaya, Malaysia

The sinusoidal wave still seems to be a matter of conjecture – Ed

Mexican free-tailed bats emerging from caves in the southern US fly in the same circular pattern, and around the British Isles puffins approaching their burrows also fly in a 'wheel flock'. Fast-flying bats or birds adopt this formation as a protection against predators such as hawks, gulls or skuas.

I am unsure about the sinusoidal movement. Perhaps it

arises because bats are safer from predators near the ground but have to seek their food higher in the sky. This could mean that sections of the flock briefly climb high in food-gathering sorties – while it is still light enough for the hawks to see them – before ducking back to low altitude for safety. Mexican free-tailed bats, for example, feed very high in the air after darkness has descended.

Sean Neill
University of Warwick
Coventry, UK

Bat sonar is oriented directly ahead of the bat and would be blocked if they flew in a straight line. By flying in a spiral each can maintain a forward view and use its sonar to detect prey or predators. Seen from behind (as in the photo) the spiral appears as a circle and from the side as a sinusoidal wave. It is remark-able how perfect the spiral is and it would be interesting to know if it is always clockwise or anticlockwise for a given species.

Jerry Whitman
Barnham, West Sussex, UK

❓ Froggy fear

I was spring-cleaning my pond and was horrified to discover what I thought was a dead fish. On closer inspection, the fish, a golden orfe, was still alive but had a frog firmly attached to it. The frog was clasping the fish as though it was trying to mate with it. I have never seen anything like it. I also noticed that several of my other normally healthy fish seem to have sustained injuries. Is a killer frog on the loose?

Clare Dyer
London, UK

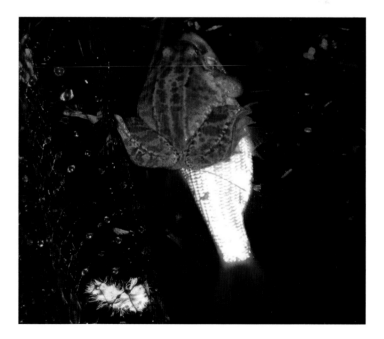

This is a case of misdirected sex drive and a lonely, rather than a killer, frog. Most frogs and toads mate in the water by means of a process called amplexus, in which the male clings to the female's back and releases his semen into the water as the eggs are laid so that fertilisation occurs externally. Males are equipped with rough 'nuptial pads' on their forelimbs which help them to grip onto the female during this process.

The strength of the male's urge to mate is dramatically demonstrated by the common toad, *Bufo bufo*, which can often be seen forming large clumps in which a single female is tightly surrounded by a large number of males. Where the sex ratio is particularly unfavourable, males will cling onto a variety of inappropriate objects, both animate and inanimate, in a futile attempt to mate.

Jonathan Wallace
Newcastle upon Tyne, UK

When the pond is overstocked with male frogs and there are not enough females to go round, the desperate males will attach to anything. If the fish is small enough it will die because the frog will cover its gills with its forelegs, thus suffocating the poor creature. I have witnessed this several times. The answer is to remove all large male frogs from your pond in early spring.

Dave Gaskell
Tranmere, Merseyside, UK

I feel compelled to suggest that perhaps what your correspondent observed is not a frog at all, but a horny toad.

Sorry, I couldn't resist.

Ben Haller
Menlo Park, California, US

❓ Shell shock

Dining out in Belgium, some of our party ate snails in garlic, and one took home an empty shell for his 3-year-old son to play with. The washed shell sat on the kitchen work surface most of the time, until one day two baby snails emerged from it. The 'parent snail' had long since been fried, scooped out and eaten. Assuming my friend is not hoaxing us, what happened here?

Dave Mitchell
By email; no postal address supplied

Several readers thought this question was indeed an elaborate hoax. But there may be a simple answer – Ed

Snails both fertilise and carry their eggs internally. When being prepared for the table, the snails are scooped out of their shells, usually mixed with butter, parsley and garlic, then cooked. After cooking they are reinserted into their shells and served.

The shells themselves are not cooked, so the baby snails that later emerged had presumably originated from eggs lodged inside the shell. These could have survived the scooping-out part of the preparation.

Gregory Sams
London, UK

The snail that is most frequently eaten throughout Europe is *Helix pomatia*, the species shown in the photograph. It is known in the UK as the 'Roman snail' because the Romans may have introduced it to these shores for food. It is native to much of Europe.

Snails that are eaten in restaurants often originate from snail farms, although they may be collected from the wild. They are hermaphrodites, but although they have both male and female reproductive organs they must mate with another snail before they lay eggs. After mating, a snail can store the sperm it received for up to a year before fertilisation, but eggs are usually laid within a few weeks of mating.

The eggs of *H. pomatia* are laid about 6 centimetres deep in holes dug in the soil. A snail will take up to two days to lay between 30 and 50 eggs. After about four weeks the fully formed baby snails hatch.

In this case, any eggs that were present during the cooking process would die. However, it is possible that the shell in which the snail was served was not the one belonging to its fried occupant. Snails are often supplied ready-cooked from producers with a separate supply of shells in which they can be placed for presentation, one of which, in this case, may have held eggs previously deposited in the upper whorls of the empty shell.

More information about snails and their conservation may be found on the website of the Conchological Society of Great Britain and Ireland at www.conchsoc.org.

Peter Topley
Bedfordshire, UK

❓ Floundering about

How do certain animals, such as the flounder, change their colour to match their background? More specifically, if you made a tiny blindfold for the flounder, would it still be able to match its surroundings?

Nick Axworthy
By email; no postal address supplied

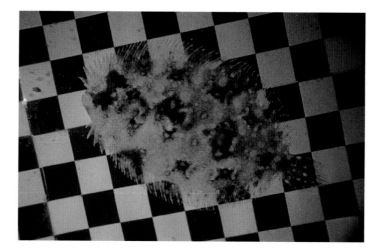

Many fish in the teleost group, such as the minnow, change colour in response to the overall reflectivity of their background. Light reaching their retina from above is compared in the brain with that reflected from the background below.

The interpretation is transmitted to the skin pigment cells, called chromatophores, via adrenergic nerves, which control pigment movement. Teleost skin contains pigment cells of different colours: melanophores (black), erythrophores (red), xanthophores (yellow) and iridophores (iridescent). Pigment granules disperse from the centre of the cell outwards and the area covered by the pigment at any specific time determines that cell's contribution to the skin tone.

Many flatfish, including flounder, go further than overall reflectivity and develop skin patterns according to the light and dark divisions of their background (as in the questioner's photograph). This seems to involve a more discriminating visual interpretation and produces distinct areas of skin with predominantly, but not exclusively, one type of pigment cell. For example, black patches contain mainly melanophores and light patches mainly iridophores, which can produce the chequerboard appearance seen in the picture.

As these responses are visual, blindfolding the fish would result in all the components of the chromatic system being stimulated equally. The fish would adopt an intermediate dark or grey skin tone similar to that on a dark night. Over time, direct light stimulation of the pineal gland through the skull will affect the amount of pigment and number of cells – a process mediated by hormones – hence the 'black' plaice sometimes sold in the UK, which have come from the sea around the dark volcanic seabed off Iceland.

Cliff Collis
London, UK

Many animals change the shade or even colour of their skin in response to certain stimuli. In cephalopods such as the cuttlefish, pigment-filled sacs can become extended (flattened) by the action of radially arranged muscle fibres that are controlled by the nervous system. Colour change in these animals is both rapid and spectacular.

In crustaceans and many fish, amphibians and reptiles, specialised pigment-storing cells in the skin called chromatophores redistribute the pigment inside them. The pigment in these chromatophores is either concentrated in the centre, or dispersed throughout the whole cell.

Imagine a white floor with a small pot of black paint standing in the middle. From above, the floor will look very light, despite a substantial amount of pigment seen as a small black spot in the middle. When the same paint is spread over the floor, the floor looks black. The beautiful trick of the black chromatophores (known as melanophores) is that they can reverse the process, effectively putting the paint back in the pot.

Flatfish, such as plaice, flounder and others, are expert at imitating not only the general shade of the surface on which they rest, but also patterns of dark and light material. Not surprisingly, perhaps, their eyes are used to perceive the shade and patterns. Light hitting the retina from above affects the ventral or lower area of the retina, while light reflected from the bottom strikes the dorsal or upper retinal surface.

If the light intensities from the two areas are similar, a signal causes the pigment of the melanophores to be concentrated in the centre of the cell, so the fish turns pale. On the other hand, when the bottom is dark the two areas of the retina receive very different light intensities, and the reverse of the signal causes pigment dispersion and a dark fish. The masters of disguise, the flatfish, can also discern patterns in the bottom surface and imitate them by regulating nerve activity to groups of melanophores.

Stefan Nilsson
Gothenburg University, Sweden

❓ Don't call me ginger

Why are orangutans orange? It doesn't seem to be a camouflage mechanism. And why are they so hairy? They live in tropical forests after all.

Peter Webb
London, UK

Orangutans' colouring does help them blend in. The water in peat-swamp forests, where orangutans live, tends to be a muddy orange. Sunlight reflected off this water can give the forest an orange cast, making orangutans surprisingly hard to see in dappled light. Many orangutan nests, up in the forest canopy, contain orangey-brown dead leaves, and some trees have reddish leaves, especially when young.

Ground-based predators would see orangutans in the canopy as a mere silhouette. In such circumstances orange may stand out less than black, which may be more suited to blending in with the forest floor. Dark African apes such as gorillas spend much more time on the ground than orangutans, while some other canopy-dwelling primates have a similar ruddy colour to orangutans. Among these are red langurs, which live in the same Borneo forests as orangutans.

As for orangutans being hairy, there are numerous possible reasons for this. Orangutans are exposed to direct sunlight up in the canopy, so hair could serve to protect their skin from the sun. It may also provide insulation and temperature control, trapping a layer of relatively cool air close to the skin by day and keeping the skin dry and warm at night and in cool rainy weather. Hair also protects against insect bites and helps break the outline of the animal's silhouette in the canopy when viewed from below.

Finally, dominant 'flanged' male orangutans have long hair on their arms and at the base of their back. This makes them look larger, helping them dominate other males and attract females.

Mark Harrison
Orangutan Tropical Peatland Project

David Chivers
Reader in Primate Biology,
University of Cambridge, UK

Orangutans may not need camouflage, given their size and strength. And if their hair is not for camouflage, the fact that reddish hair is common in primates such as proboscis and red leaf monkeys, and in other mammals such as tree shrews, squirrels, foxes, deer and flying foxes, suggests that it isn't an awkward thing to have.

As to whether being orange provides camouflage, that depends on the environment. Some marine species that live at depth are well camouflaged despite being bright red, because red doesn't stand out in the low light under water. Something similar applies to orangutans.

The explanation lies in the way sunlight penetrates the forest canopy, bouncing off vegetation as it does so. Leaves absorb red, orange and violet light for photosynthesis, reflecting green. So by the time sunlight has reached the forest floor, it has been robbed of its reds and oranges. In this sort of light, orangutans look like large, dull brown lumps. They can be so well camouflaged that several times I have walked past having failed to see them sitting on the ground half a metre away. I have twice come within a hair's breadth of tripping over them.

The explanation above is something I heard from science writer and evolutionary biologist Jared Diamond, who has studied birds of paradise in Papua New Guinea. He described males displaying in columns of dappled sunlight: as they moved in and out of the light, they changed from dull brown to spun gold, as though disco dancing under strobe lights.

Anne Russon
York University
Toronto, Canada

The writer is the editor of The Evolution of Thought: Evolution of great ape intelligence *(Cambridge University Press, 2007) and the author of* Orangutans: Wizards of the Rainforest *(Firefly, 2004) – Ed.*

Several replies noted the similarity between the words 'orang' and 'orange' – but only one explained their derivation in detail – Ed.

Orangutans may be orange, but the name has nothing to do with their colour. It comes from Malay and means 'person of the forest', *orang* being Malay for 'person'. The word 'orange' – originally meaning the fruit, but later used for the colour too – came into English via a trail of other languages.

It ultimately comes from India (Tamil via Sanskrit), from where it made its way into English in the 15th century via Persian, Arabic, Italian and French. So 'orangutan' and 'orange' are unconnected etymologically.

Linguistic coincidences are quite common, as it happens.

The word for 'dog' in Mbabaram, an indigenous Australian language, is *dog*, although the word is not borrowed from English. Similarly, the word for 'honey' in both Hawaiian and ancient Greek is *meli*. Chance similarities are a major problem in establishing whether two languages derive from a common ancestor.

Sometimes these coincidences take on a life of their own. At one time a berfry was a tower, but when people started to associate the first part of it with the word 'bell', it changed to 'belfry' and started to mean a tower with a bell in it. So far, no one is claiming that orangutans are so named because they are orange, but it's not inconceivable that one day someone might.

David Willis
Department of Linguistics
University of Cambridge, UK

❓ Catching the red eye

Some bird species, such as the great-crested grebe, hunt underwater for fish and have red eyes. The red colouring is presumably beneficial to these diving birds, but in what way? If it does provide an advantage, why have other birds with similar habits not evolved red eyes?

Ian McKechnie
Weybridge, Surrey, UK

The red eye is caused by the colour of the iris, which controls the diameter of the pupil. Only some diving birds have red irises. The eye of a king penguin, for example, is black, and set against black plumage.

The pigmentation of the iris is just a device for making the iris as opaque as possible to accurately control the amount of light reaching the retina, by ensuring no light leaks through the iris muscles. Its colour is in this sense irrelevant.

But eye colour was extensively reviewed by ophthalmologist

Ida Mann in 1931, who concluded that iris colour is used as a signal between animals. There has been no evidence to counter this conclusion since.

So that a structure used for signalling is conspicuous, it needs to contrast with its surroundings, and so in some birds – such as the great-crested grebe mentioned in the question – bright iris colours often contrast strongly with the feather or skin of the head in birds (or, in certain cases, they match it).

Eye colour varies widely. There are grebes with yellow eyes, penguins with black, red or yellow eyes, and cormorants with blue, green, red or black eyes.

Iris colour changes with age in many birds, with the young showing only browns and blacks and not attaining the full bright colour until adulthood. Presumably this is when they need to use iris colour as a signal of fitness or when emotional state becomes important.

Graham Martin
Emeritus Professor
Centre for Ornithology
School of Biosciences
University of Birmingham, UK

❓ Tiger, tiger

Why do tigers have stripes? The other big cats tend to have spots.

Linda Veron
Tarragona, Spain

The beautiful striped markings on tigers' coats are unique in the cat family. Other closely related big cats have spotty rosette or cloud-shaped body markings (leopards, jaguars and clouded leopards), or plain coats (lions).

Work by our team at the University of Bristol has shown that cat patterning evolved to provide camouflage suited to the cats' particular habitats and behaviours, enabling them to capture prey more effectively (and escape predation in the case of the smaller cats). In general, plainer species such as lions live in open environments and hunt by day, whereas cats with

complex patterning like leopards and tigers have more nocturnal habits and live in environments with more trees. Our sta-
tistical analysis nicely supports commonsense natural history.

Unfortunately, with no other striped cats around besides
tigers, we cannot use the same methods to identify the evolutionary factor which drove tigers to depart from the ancestral
big-cat pattern. Tigers are much bigger than jaguars and leopards, but in general they have a similar ecology, and tigers'
historical range and habitat overlap considerably with those of
leopards. So why don't they look similar?

One idea put forward decades ago, but for which evidence
is still lacking, is that compared with the typical leopard habitat,
the average tiger habitat contains a lot of vertical features such
as bamboo. Quantifying this would be straightforward except
that with the tigers' range now so shrunken, it is hard to know
exactly what sort of forest their coat evolved in. Tigers are obviously well camouflaged, yet the factors behind their appearance
remain an enigma.

Intriguingly, our team has also shown that big-cat patterning changes relatively rapidly over evolutionary timescales.
One day our descendants might wonder at the beauty of striped
leopards and spotty tigers.

Will Allen
University of Bristol, UK

Your earlier correspondent's discussion of the tiger's stripes was
excellent. However I must take issue with the comment '… with
no other striped cats around besides tigers …'. The tiger may
be the only large striped cat, but there are a number of smaller
striped wild cats. The European wild cat (*Felix Silvestris*) and its
subspecies are striped. The sand cat (*Felis Margarita*) is striped,

and the fishing cat (*Prionailurus viverrinus*) has both stripes and spots that vary in dominance depending on the individual cat.

David S. Rubin
By email; no postal address supplied

2 Ice, bubbles and liquid

❓ Air spray

I wanted to chill a mug of water so placed it in the freezer, but then forgot about it and it froze solid. When I removed the block of ice from the mug it contained the most amazing thistle-like pattern of what seemed like canals of air. None of these canals extended to any outside surface. What happened?

Brian Barnes
Somerset West, South Africa

In the freezer the water cools from the outside in. So the first crystals of pure ice form around the outside of the mass of water. As these ice crystals grow inwards, the air dissolved in the water becomes trapped and its concentration increases. Initially it remains dissolved, because cold liquids can hold more dissolved gas than hot ones. This is why the outer layer is almost free of air bubbles.

However, the concentration of dissolved air eventually exceeds the ability of the water to retain it in solution, so air bubbles begin to form and are trapped in the ice as the crystals grow inwards, forming the patterns observed. The lines all curve downwards because water on the verge of freezing is less dense than slightly warmer water, and so rises. Thus the water freezes from the top down, meaning that the water at the bottom (which is also somewhat insulated from the cold air in the freezer) is the last to freeze.

Simon Iveson
Department of Chemical Engineering
University of Newcastle
Callaghan, New South Wales, Australia

If the container is smooth, the outer layer is bubble-free because solutes in the water, gases in particular, are unsaturated when freezing starts. Bubbles form only after some ice has formed, forcing enough gas into the surrounding water to supersaturate it. Initially, tiny spherical bubbles appear on the advancing surface. What happens next depends on the path of the ice front and the form of the growing crystals.

In this case, the front advanced smoothly along the curves described by the long bubbles. The original spherical bubbles caught in place by the growing crystals acted as nuclei into which the rest of the gas collected as it escaped from solution. Being surrounded by ice in every direction except perpendicular to the ice surface, the bubbles formed tubes whose shapes traced the advance of the crystal-water interface.

It's common to see such bubble growth, but the lovely symmetry and consistent bubble structure shown in the photograph require still water and suitable solutes, temperature gradients and crystal forms. To sculpt such bouquets in various forms, and possibly in other media, should make for satisfying science experiments and art projects.

Jon Richfield
Somerset West, South Africa

❓ N factor

Leaving the house early one frosty morning, I noticed the emblem on my car was frosted over on all the letters but one. The N was completely frost-free as you can see in the photo. The outside temperature was hovering around 0 °C. If all the letters were constructed of the same material, surely they would all appear the same, so why is the N special?

Nikki Cherry
Peterborough, Cambridgeshire, UK

The N is frost-free because it is in poor thermal contact with the car body, perhaps because it is slightly loose or there is a small gap behind it.

All objects cool by emitting infrared radiation. Metals cool more effectively in this way than plastics because metals have a high thermal conductivity, which means it is easier for heat inside a metal object to move to the surface and escape.

As this cooling occurs, the surface of an object initially becomes cold enough for water to condense on it (take a look at a cold beer bottle on a humid summer day). When the ambient temperature is close to freezing, the surface may eventually get cold enough for that condensed water to freeze, or for water vapour to condense directly onto it as small ice crystals. Both of these processes give rise to frost.

The letters in the manufacturer's logo are made of plastic. If they were somehow floating just above the car body, they would all stay frost-free because plastic cannot radiate heat fast enough for them to frost over. It is because they are in contact with the car body that they can lose additional heat to the metal via conduction, allowing them to get cold enough for frost to form.

The N must be in poor contact with the metal for some reason, hence the absence of frost, though you can just make out it has become cold enough for water to condense onto it.

Adam Micolich
School of Physics
University of New South Wales
Sydney, Australia

⍰ Cranberry ice

One of my faculty colleagues, Michael Runtz, took this photo of ice bubbles in Cranberry Lake in Ontario. How did the bubbles form in this amazing fashion?

James Cheetham
Department of Biology
Carleton University
Ottawa, Ontario, Canada

Without any scale reference or indication of the depth of the lake I cannot tell for certain, but I have seen similar bubbles frozen in ponds where I grew up in upstate New York.

In freshwater ponds and lakes, the biological activity of microbes in the sediments on the lake floor produces bubbles of gas, usually methane or carbon dioxide. In winter this activity is slow, but it is still present.

The gas bubbles rise to the frozen surface of the lake, becoming trapped there. The following night, another layer of ice forms beneath the bubble, so it is encased in ice. This leads to the flattened shape you see. The picture is a frozen daily record of the gas emissions.

Obviously you need calm, shallow-water ponds or sheltered water at the edges of small lakes in order for this phenomenon to occur.

James Field
Aberystwyth, Ceredigion, UK

? Crossed channels

While cycling in Ireland I had ample opportunity to observe rain and puddles, and I took this photo of muddy water running across a road. Why has it separated into bands, and what determines their spacing?

Robert Johnstone
Leeds, UK

What you see here is the effect that changing the velocity of a fluid has on any solids being transported by it. If the fluid is moving with a velocity above that at which the solid would normally settle out, localised currents in the fluid will stop the solid settling normally. I suspect the road has some surface imperfection which caused an initial drop in velocity, allowing some of the sediment load to settle onto the road surface as a band. The second band is then caused by the drop in velocity generated by the first band and so on, with the spacing determined by the gradient and roughness of the surface. The bands are curved because the water eventually builds enough pressure to force its way around the sides of each band of sediment. With this pressure comes velocity and so, once the velocity is high enough, no more sediment settles out in that band.

The same principle of solids settling because of a drop in velocity is used to separate minerals such as gold from river sediments using a pan.

Neil Ayre
Kalgoorlie, Western Australia

This is a result of a complex series of interactions involving, among other factors, the speed of the fluid flow, the viscosity of the fluid, and the ratio of the thickness of the top layer to the overall thickness of the fluid.

This and related problems were first studied by John Scott Russell in the early 19th century. In the later part of that century Diederik Korteweg and Gustav de Vries modelled the phenomenon with an equation that bears their names. The types of self-reinforcing waves formed in this situation are known as solitons and they are fundamentally different from normal waves occurring in bodies of fluid such as the sea, where the depth of the fluid is very large compared with the height of the wave crest.

Doug Dean
Pfeffingen, Switzerland

This occurs when water or other fluids run in a thin film over a surface. Male readers may sometimes see the effect when using urinals – Ed.

The waves shown in the photograph are a classic example of what are known as roll waves. Other examples include the 'urinal effect' noted by your (male) editor and also rainwater flowing down a window (which is effectively the same thing). On a larger scale, roll waves sometimes appear on spillways discharging the overflow from reservoirs, where they can become large enough to overtop the sides of the channel that would comfortably contain a uniform, steady flow. In extreme cases the flow effectively moves in surges with very little water in between, which gives the phenomenon another name: slug flow. Roll waves occur on the free surface of liquids, liquids carrying solid particles in suspension, slurries, and also at the interface of immiscible liquids such as oil on water.

Contrary to what your earlier correspondent suggests, they are quite different from solitons, which are essentially discrete pulses of liquid moving on top of the otherwise undisturbed flow beneath them, and nor are they miniature waterfalls over static bands of silt. If these existed, they would obscure the line in the road running from top right to bottom left of the photograph.

Roll waves have been studied for more than 80 years. For a gentle slope the flow will be slow and deep – what is known as subcritical flow – while for a steep slope the flow will be shallow and fast-moving – or supercritical. The difference between these two states is given by the Froude number, named after the 19th-century fluid dynamicist William Froude. This number is the ratio of the velocity of the flow to the speed of very small waves that invariably appear on the liquid's surface, and can be greater than or less than 1. For supercritical flow it is greater than 1, which means that the waves move slower than the flow and therefore can only travel downstream.

The speed of waves increases with their height, so larger waves will overtake and absorb the smaller ones (which also increases the speed of the large ones). Gradually, the many tiny waves become fewer, larger ones. Eventually the flow in their vicinity becomes subcritical, the wave fronts steepen and they break in much the same way as waves break on a beach.

Roll waves do not appear at regular intervals and there is no way of calculating the average distance between them. They appear spontaneously even when the flow is over a smooth surface, provided the Froude number is greater than 2. A slightly rough surface appears to promote their appearance, but further increasing the roughness has the opposite effect and ultimately will prevent them occurring altogether.

Richard Holroyd
Cambridge, UK

❓ Frozen images

In January I dropped some bricks into my pond, which is a metre deep. In March the pond froze over and an image of the bricks appeared like a hologram in the ice – as shown in the photo. What caused this?

Clive Gardner
Glenfield, Leicestershire, UK

Pond water contains a certain amount of dissolved gas, including oxygen. Because of the physical properties of water, the colder it is the less gas per unit volume it can hold. Water is at its densest at a temperature of about 4 °C.

As the water temperature in the pond drops, cooled by the colder air above, the surface water sinks to the bottom by convection. Once the whole body is at 4 °C this convection stops because further cooling makes the surface water less dense. The surface starts to freeze and the coldest water begins to release its dissolved gas. Some of this would bubble upwards, but much more would diffuse down, still remaining in solution, until eventually the water surrounding the bricks becomes supersaturated.

The rough surface of the bricks, particularly around the edges and corners, provides nucleation sites for dissolved gases. Gas molecules collect preferentially around the edges of the bricks, eventually producing bubbles. As these reach a critical size they break away and float straight upwards in the still water. Because there is a layer of ice on the surface, the bubbles become trapped and frozen into it. As the ice layer thickens and bubbles continue to rise from the brick, the 3D shape develops. The rate of bubbling was probably very slow, as was the rate of freezing, which allowed the very detailed effect to form.

David Jackson
Liverpool, UK

A previous correspondent states: 'Because of the physical properties of water, the colder it is the less gas per unit volume it can hold.' Actually, the opposite is true: water holds more gas as it gets colder. This is why opening a bottle of fizzy drink on a hot summer day releases more fizz than opening one straight from the fridge.

What does happen is that water releases most of its dissolved gas when it freezes. In the case of this question, the layer of water on top of the pond releases bubbles of gas when it freezes, which is why the water below becomes supersaturated with gas.

Tim Patru
By email; no postal address supplied

❓ Ice vines

Living in an older home with single-glazed windows, I have grown used to seeing intricate patterns of frost on the panes each winter. However, I was truly impressed by this twining, vine-like pattern that appeared one January, and I would love to know how it came about. The vine-like shapes formed in a 20-by-30-centimetre section of the window and were surrounded by standard snowflake-shaped frost. The 'vines' were 1 centimetre wide with small dots running up the centre, and they twisted about each other with leaf-like shapes sprouting from the sides. The photo below shows a section measuring about 6 by 10 centimetres. It had been a particularly cold day (−20 °C) and the sun was shining on the window. My wife suggested that the sunlight shining through the branches of a tree 2 metres away had caused this, but there were no distinct shadows visible on the window at the time.

Ken Zwick
Neenah, Wisconsin, US

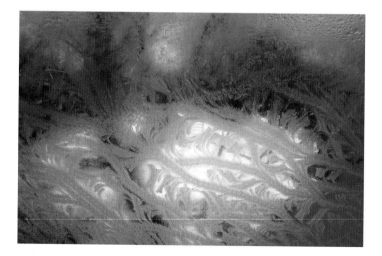

No clear answer to this one, but some hypotheses. This phenomenon has been seen by many people – Ed.

During the winter my conservatory roof is regularly covered with an even layer of frost crystals in no particular pattern, but on three occasions the entire 20 square metres of double-glazed glass has been covered with 'ice vines' (see photo below). Because there are no trees anywhere near my conservatory the patterns could not have been formed by sunlight shining through branches.

The ice vines seem most likely to form on frosty evenings after mild, sunny winter days when air pressure is high, the sky is clear and there is a gentle breeze. At these times, bands of condensation form on the glass with dry stretches between them.

I assume that the breeze flowing over the conservatory frame, which stands about 4 centimetres above the glass, sets up an oscillation in the air, rather like the waves produced in a wind instrument such as a recorder or a whistle. As the air touches the glass, chills it and bounces off again, this may form the bands, which always run parallel to the frame, although the width of the bands varies with wind speed. Generally, the 'stems' of the ice vines follow the direction of the frame, so it appears that the bands of condensation and the ice vines are somehow linked.

When growing crystals, the finest and largest specimens form when the crystals grow very slowly. It is possible that the gently oscillating air over the conservatory roof reduces the rate at which water molecules crystallise, adding to the complexity of the ice patterns. Also, as the breeze dies away during the night, cold air sliding down the slope of the roof may change the direction of the oscillations, causing even more intricate patterns.

Steve Antczak
Lymington, Hampshire, UK

I have a picture of an almost identical pattern on my car's rear windscreen (see below) after it was parked overnight away from any trees during frosty weather.

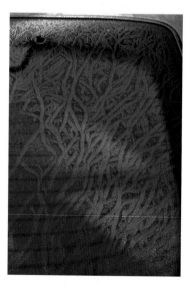

It is a very striking pattern. It looks as though something has progressed in a stuttering way along the vines. I have had the same car for several winters but have spotted this effect only once.

Alan Singlehurst
Shildon, Durham, UK

We received further photographs (one is shown below) of this astonishing phenomenon from reader Steve Redpath of Aberdeenshire, UK, who spotted ice vines on his roof window last winter – Ed.

❓ Snow squares

This photograph was taken mid-morning after a light fall of crisp snow in the night. The temperature had not risen much above 0 °C. I have always assumed that the circles form because the concrete slabs conduct heat differentially from the open gaps between them. But what causes the centres to melt?

Neil Howlett
Frome, Somerset, UK

Working on the assumption that 150 correspondents can't be wrong, we selected the three following answers – Ed.

The pattern is caused by the five-blobs method of slab laying, in which a blob of mortar is placed under each corner of the slab, and one at the centre. This makes it easier to level the slab. These blobs are conducting heat from the ground underneath, melting the snow in this pattern.

Steve Law
Kingston, West Sussex, UK

Regular viewers of TV gardening programmes will know that the preferred method for laying a patio is to lay the slabs on a full bed of mortar, so that the slab is supported across its entire undersurface. Laying on a full bed will not only stop this pattern showing up, but will also ensure that ants don't make a home in the gaps under your patio slabs.

Gary Stanley
Castleford, West Yorkshire, UK

With a bit of foresight, a creative builder could produce an effect whereby interesting patterns, pictures or words could appear as a light snowfall melts.

David Turner
Sevenoaks, Kent, UK

❓ Ring of bright air

While watching beluga whales at Vancouver aquarium I noticed that one of them was blowing air rings. These appeared as annular bubbles that exhibited no obvious buoyancy and could be propelled to the bottom of the pool. Some of the whales would blow one horizontally then swim over to it and suck it back in. Have there been any studies of these intriguing air rings? How fast can these bubbles be projected and to what depth and does it vary between animals? And why do they produce this shape – does it have any beneficial purpose?

John Chapman
North Perth, Western Australia

Unlike soap bubbles in air, air bubbles under water are maintained by the pressure of the surrounding water rather than tension. The beautiful bubble rings blown by dolphins are actually toroidal vortices that they create by sending a localised gush of water into an otherwise still region, which they then inject with air. This air moves to the centre of the vortex because that is where the pressure is lowest.

Stir water rapidly in a cylindrical glass with a smooth rod (to avoid too much turbulence) and you may see the centre of the vortex sharply depressed where the pressure is lowest. If you stir faster, bubbles may detach and move down the core of the vortex, illustrating the behaviour mentioned above.

Steve Gisselbrecht
Boston, Massachusetts, US

A ring-shaped bubble is an example of a vortex ring, which is similar to a columnar or tornado-like vortex, but bent round into a circle. Other examples include smoke rings. They are formed when a flow through a circular opening is forced back on itself.

There is also a second class of bubble ring. These consist of water vapour in water, and form when a vortex ring generates low enough pressure to cause vapour to form around the ring of the vortex, giving it a similar appearance to an air ring under water, though filled with vapour. This type of 'cavitating' vortex is used by some underwater weapons systems.

David Hambling
London, UK

Cavitation vortices can be seen spinning off the tip of ships' propellers, and they form when the liquid flows so rapidly that its pressure drops enough for it to turn into vapour – in essence the water boils underwater.

There is another even more exotic way to create such bubbles. Reader David Williamson of London points out it is possible to create 'antibubbles'. An antibubble is a bubble in reverse. Just as the soap bubbles that children blow are a thin skin of liquid in air, an antibubble consists of a thin skin of air with water inside and out.

Their properties have been explored by S. Dorbolo, H. Caps and

N. *Vandewalle of the Institute of Physics at the University of Liège in* Belgium, *and described in a paper published in 2003 in* New Journal of Physics. *You can find out how to make antibubbles at www.anti-bubble.org.*

Alex Vallat of Cambridge, UK, tells us that blowing air rings underwater is not difficult to do. Puff your cheeks out with your lips pursed. Then, with your throat closed, make a P sound with the lips and use the stored air to blow out quickly (see photo on previous page). The tube that forms the bubble rotates around its core, like the smoke rolling around a smoke ring, but the ring itself does not spin around its centre like a steering wheel.

Steve Backshall of Wooburn Green, Buckinghamshire, UK, has a different approach. Try sitting on the bottom in relatively still water, he says. Then rock backwards so your mouth is facing upwards, place your tongue firmly on your upper lip, then forcefully expel air before briefly sucking back in and closing your lips. Accomplished ring blowers can create mesmerising, expanding doughnuts of shimmering air that gyrate towards the surface (see below).

Dolphins blow toroidal rings for fun and then play with them underwater. Reader Alistair Eberst of the University of Abertay Dundee in Scotland drew our attention to www.earthtrust.org/del-rings.html, where you can see pictures similar to the one below, and much more – Ed.

In 2011 The first 'King of the Bubble' contest was held in a specially built 33-metre deep pool in Brussels, Belgium. Contestants had to produce the largest, clearest, longest-lasting, perfectly rounded ring bubbles in order to take victory.

Peter Mann
Ruislip, Middlesex, UK

One answer sometimes leads to another question. We have heard numerous anecdotes of dolphins, whales and even human swimmers who have been able to create stable toroidal bubbles capable of existing underwater. And the questioner below wants to know if they can be created in any environment. Do any of our readers know what the secret to these bubbles is? – Ed.

❓ Soap on a hope

Is it possible to blow a toroidal soap bubble (one shaped like a ring doughnut)? And if it is, would it collapse immediately to a sphere? Could its life be prolonged by spinning its surface, as with smoke rings?

Peter Gardner,
Blawith, Cumbria, UK

A soap bubble is the minimum surface which encloses a given volume. If a toroidal bubble were created, it would not provide such a minimum surface and would therefore tend to contract to reduce its surface area until it collapsed into a bubble which would then burst because of the forces created at the disappearing hole in the torus. This situation differs from that in a solid torus such as a bicycle inner tube, because soap bubbles can transfer part of their surface from the inner to the outer part of the torus as they shrink.

A temporary toroidal bubble could perhaps be created by sticking spherical bubbles in a ring and collapsing their shared walls, but the inner ring would undoubtedly degenerate as the number of bubbles decreased.

Soap bubbles are different from smoke rings, which have no surface but are composed of solid particles suspended in air. These are stable because different parts of the body can rotate at different speeds without causing degeneration.

Jerry Humphreys
Bristol, UK

As a mathematician who studies soap bubbles, I knew that a toroidal soap bubble was, under normal circumstances, impossible. The only stable equilibrium shape for a soap bubble is the sphere that most people easily recognise – a torus bubble should not even exist in unstable equilibrium.

So when the famous performer Tom Noddy (known as the Bubble Guy from the US TV show *Tonight*) told me that he once blew a toroidal bubble, I didn't actually believe him until he showed me the photographic proof (below). The bubble didn't last long, but it did exist briefly. Visit www.tomnoddy.com to see some further interesting examples.

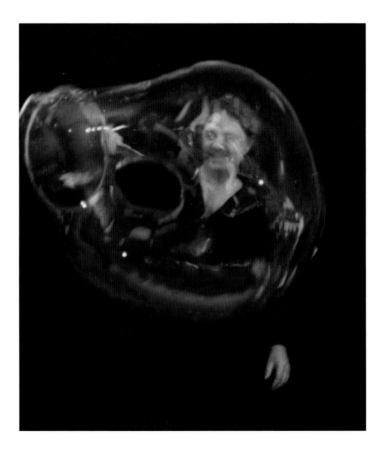

Torus bubbles do occur in unstable equilibrium in double soap bubbles: an outer bubble wrapped around another at the centre, as in the diagram below – a copy of a computer simulation created by John M. Sullivan, Professor of Mathematics at the University of Illinois. More of his images are online at http://torus.math.uiuc.edu/jms/images/.

Frank Morgan
Williams College
Massachusetts, US

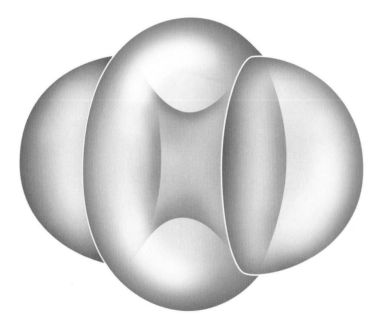

? Woodland wonder

The ice crystals in the photo were found on small branches lying on the ground in mixed woodland. The crystals had formed only where the bark was missing and there was very little frost elsewhere on the ground. Can anybody explain how they formed?

David Meadows,
Yeovil, Somerset, UK

The explanation behind this photograph of fine, frilly ice filaments growing from small branches is that the hairs of ice are generated from water held in the pores of the decaying wood, which is sucked out by the freezing action to form filaments. For this to happen there would have to be no free water on the outside of the wood. Indeed, the questioner notes that there was little frost elsewhere on the ground.

I have seen the same ice crystals on rotting elder wood alongside the river Barle, downstream of Wimbleball dam, also in Somerset.

In the example in your questioner's photograph the moisture held in the pores of the wood was still liquid and was being extruded and frozen at the surface. The fibres reflect the size and spacing of the pores in the substrate. Their curvature was caused by the weight and drag at the point of freezing (like meat coming out of a mincer). When the temperature differential is low, ice crystals are not able to penetrate fine pores and are extruded instead. Water will even be extruded as ice crystals from between the minute particles in glutinous sludges, leaving a more crumbly and concentrated residue. This was the basis of a process once used for treating waterworks sludges.

A more common phenomenon is columnar ice, which can be seen on chalk downs, for example, where a layer of ice grows out of damp porous lumps of chalk in frosty conditions.

David Stevenson
Newbury, Berkshire, UK

Miracles have no explanation, but magic does and, in this case, bottles of milk left on an icy doorstep provide the model. On two mornings this winter I found some 10 centimetres of frozen milk extruded from my pint, curved to some extent and still wearing the aluminium cap. The remaining liquid was in the form of a slush. I should add that this effect was very common 50 years ago before global warming set in.

For this phenomenon to occur on soil and rock requires a source of water held in pores to cool gradually from 4 °C to 0 °C, or perhaps lower under pressure. As some of the water molecules begin to form short-range crystalline structures, the water expands. Under these conditions, ice fibres are extruded through the pores. Gravity and the phenomenon of regelation – when partially thawed ice re-freezes – may cause the curvature. The ambient conditions described by your correspondent were perfect.

Philip Sutton
Gateshead, Tyne and Wear, UK

? Wrapped bubble

I found this strange narrow spiral bubble trapped in a frozen puddle in Weardale, County Durham, UK. Nearby puddles of about the same size had typically a dozen or so irregular sausage-shaped bubbles, usually wider than this spiral but displaying a similar concentric nature to varying degrees. The one photographed was the most striking, however. How and why did it form?

Bob Johnson
Durham, UK

In this case, dissolved gas is not the issue. The 'bubbles' are the result of the puddles sitting on porous ground.

As the puddle begins to freeze, the water is still draining away under the ice. If the ice cover is thin enough the unsupported surface sags. A meniscus can form between the ice cover and the water below, and this then freezes to form a bridge to the ground.

The freezing tends to draw water up from the ground through these menisci because they conduct heat better than the air of the bubbles in between. The result of this process is usually a series of large, flat, frozen air spaces, but in this case there has been an unusually long progression of a freezing meniscus that has followed the line of contact with the receding water under the concave ice cover.

The rate of freezing and the rate of drainage of the water must have been just right for this spectacular spiral figure to form.

David Stevenson
Newbury, Berkshire, UK

? Tank trap

The Plum Temple in Zhaoqing, near Canton, in China, features an optical illusion tank (below) about 4 metres long and a little under a metre across. The puzzle is in the perceived depth. If you stand at the right-hand end, the water appears to be about 1 metre deep, sloping away to the left end where it appears to be about 10 centimetres deep. If you move to the left end and look to the right, then the appearance is reversed and you are now at the deep end. If you stand in the middle, the bottom appears to be U-shaped, with the deepest point exactly in front of you, sloping away to become shallow at both the left and right ends. When I visited, not even the technical types in our party could explain it. The temple was built in the 10th century, so this tank has been baffling visitors for more than 1000 years. Can anyone offer any explanation?

Phillip Bruce
Hong Kong

The optical illusion shown in the reader's photograph is simply a consequence of the bending, or refraction, of light as it leaves the water in the tank.

Light travels at different speeds depending on the medium it is in, and it travels faster in air than in water. The change in speed causes a ray of light to be bent where the water meets the air. Assuming a flat water surface, a ray of light coming from beneath the water at the far end of the tank will be closer to being parallel to the water surface than a ray coming from directly underneath the observer. The nearer the ray is to being parallel to the water surface, the more it bends towards the observer. But the observer's eyes do not account for the bending and assume that the light is travelling in a straight line. This makes it seem as if the ray is coming from a point that is shallower than the true point of origin.

Therefore, the closer part of the tank's floor will appear to be roughly at its 'true' depth, whereas the far end of the tank will be seen as shallower.

Using computer ray-tracing software, I have recreated the scene as best I can (see photographs opposite), using the correct optical properties of water. You can see the view from the right-hand end of the tank and the view from the middle of the tank. These images were rendered using a free ray-tracing program called POV-Ray (www.povray.org). This program takes a description of a scene, written in its own peculiar language, and draws it by tracing the path of light from its final position in the image back onto the objects in the scene, while taking into account effects such as reflection, refraction, fogging, filtering and emission. This is computationally intensive, and despite the simplicity of this scene, both images took over two minutes to render on my ageing 233-megahertz Pentium-II PC.

Beginning with the estimated size of the tank given by Phillip Bruce, I constructed a hollowed-out box for the tank, and placed another smaller box inside it. The smaller box was given the optical properties of water (95 per cent transmission, index of refraction of 1.33), except that I reduced the amount of reflection in order to show the tank floor more clearly. An infinite

plane creates the floor, and another box creates the back wall. A soft light was added. POV-Ray is unable to produce brick-like textures, so I gave it a chequered pattern instead. Many minor adjustments were needed to get the cameras into suitable position.

Andrew McRae
Queensland, Australia

Refraction is part of the explanation. The bending of light at the water surface makes the tank's flat bottom look curved. However, this does not explain the fact that the cement lines in the tank wall's tiling are straight.

In fact, these tiles are not under the water but are the reflection of the wall above the tank off the water's surface. The tank walls are dark, probably the same colour as the wall above, but they are obscured by the reflection of the upper wall. Hence you see the reflected upper wall with straight cement lines, as these lines are not influenced by refraction.

The base of the tank is light coloured so that it reflects more light than the upper wall surface reflection, and hence we see the brighter image – the principle of the one-way mirror. If your enquirer were to return to China and paint the bottom of the tank dark grey, the illusion would be lost and only the reflection of the back wall and the signs would be visible.

Doug White
Dogmersfield, Hampshire, UK

The illusion is a more extreme form of the optical distortion present in any swimming pool, and is strengthened by the observer (like your photographer) having his eye closer to the water surface. It can also be strongly reinforced if the refractive index of the liquid in the tank is increased, as will happen if its density is increased by adding solutes such as salt.

The dense solution could be overlaid with fresh water to yield a stepped density gradient which would increase the illusion. Such gradients resist thermal convection, and are stable enough to be used in Israel for solar collection ponds near the Dead Sea.

I wonder if the mottling of stone and brick is because of the presence of salts or other solutes in the tank.

J. O. N. Hinckley
London, UK

❓ Frozen veg

I took this picture (below) of ice in a field earlier this year. Individual grass stalks are encased in the ice. Nearby was an irrigation system that emitted a fine spray at certain times of day. Are the two connected?

Liz O'Neill
By email; no postal address supplied

Your correspondent is correct in supposing that the fine spray of the irrigation system produced the near-vertical icicles on the grass blades. The photo on the next page shows a similar phenomenon. This one was taken at an industrial plant in south Wales in 1953. The leaking fire hydrant on the left of the picture produced a fine spray which froze into icicles on the nearby brick wall. Odd that I should wait 60 years before finding a use for the photo...

David Gregorie
Waikanae, New Zealand

The two are certainly connected as I saw a similar phenomenon in Iceland earlier this year. The Gulfoss waterfall was throwing up a fine, cold mist of spray which condensed on blades of grass and then froze in the cold air. This created ice-enclosed grass similar to that described.

Charles Harrison
Portsmouth, Hampshire, UK

I took this photo (opposite, at top) in a field near my house a couple of years ago. The pipe running into the water trough was emitting a continuous fine spray. I presume the reason such fine blades of grass remain standing with such a weight of ice on them is due to the fact that the ice was deposited in many thin layers.

Sophie Yauner
Albury, Surrey, UK

Undoubtedly the irrigation system was responsible. A few years ago I took this picture (opposite, at bottom) of a water

plant in my outdoor pond. It was close to a fountain and repeat-edly splashed with water during a period of sharp frost. As you can see, the water froze onto the plant stems and formed these amazing, golf ball-sized spheres.

Barry Soden
Bexhill-on-Sea, East Sussex, UK

Because a limited area is affected, the sprayer is almost certainly responsible. The ice is so clear because fine droplets formed delicate films, first on the individual blades, then on wet ice barely below 0 °C. The freezing water film is so thin that gases coming out of solution escape instead of forming misty bubbles in the ice. Water supply rate, temperature and humidity are too high for droplets to freeze while still in the air, or for hoarfrost crystals to form, so we get those clear individual ice lollies.

A similar effect is important in protecting frost-sensitive fruit trees in places like the South African Great Karoo. When night temperature and humidity in still winter air drop below critical levels, then farmers use sprayers. This prevents the dreaded dry 'black frost' that forms at temperatures below 0 °C destroying the crop for the coming season. Instead, the spray freezes gently into a protective layer of relatively 'warm' ice on the buds and twigs. Latent heat released when water freezes keeps the temperature close to 0 °C.

Jon Richfield
Somerset West, South Africa

3 Clouds and stuff in the sky

❓ Clouding the issue

On holiday in Taormina, Sicily, about 30 kilometres north-east of Mount Etna, which we could see from our window, we awoke at 6.45 am to see an odd cloud drifting towards us. What caused it? Was Etna responsible? My grandson suggests it's a flying saucer, but I'm sure that's not it.

Joyce Lowe
Newtown Linford, Leicestershire, UK

The clouds shown are lenticulars, which are caused by waves in the air downwind of mountain ranges. Lenticulars do not drift, but form continuously as moist rising air from the upwind side of the mountain condenses. As the air descends on the downwind side it warms and the cloud evaporates. Any observed movement of the cloud is in fact caused by a change in wavelength.

Solitary mountains rarely produce waves strong enough to form clouds, but in this case airstreams deflected around each side of Mount Etna may meet on the downwind side and contribute to the updraft. Lenticulars are often best seen in the early morning before thermals that form during the day disrupt the wave system.

Andrew Brown
Glider pilot
London, UK

The formation is a lenticular cloud, or to give it its technical name, an altocumulus standing lenticularis, and is almost certainly connected with nearby Mount Etna. This is not because it is a volcano, but simply because it is a high mountain close to the sea. When stable moist air, such as wind blowing off the warm Mediterranean sea, flows over mountains, a series of large-scale standing waves can form on the leeward side, and lenticular clouds can form at their crests.

Not only are lenticular clouds striking to look at, they also provide useful signposts for aviators, albeit for quite contradictory reasons. Pilots of large aircraft try to avoid lenticular clouds because of the threat posed by the extremely powerful rotor forces that fashion their distinctive shape. Glider pilots, on the other hand, will actively seek out 'lennies' to use those same vertical air movements to obtain lift. Indeed the current altitude and distance records for gliders were set employing this so-called 'wave lift'.

It has been claimed that the phenomenon played a decisive

role in the 1942 battle of the Coral Sea, which lies between Australia, Papua New Guinea and the Solomon Islands. A force of elderly US Navy Devastator torpedo bombers, hunting the Japanese fleet and low on fuel, found their path blocked by Papua New Guinea's Owen Stanley mountains. At almost the point of no return Commander W. B. Ault, the formation's leader and an experienced glider pilot, identified the distinctive cloud patterns associated with wave lift and used them to enable his squadron to soar over the range, where they found and sank the carrier *Shoho*, the first major Japanese warship sunk during the Second World War, in what proved a turning point in the Pacific war.

Your reader's grandson is not the first to propose 'flying saucers'. Veteran UFO debunker Donald Menzel, and the US government's Condon report – which ruled out the existence of UFOs back in 1968 – have pointed out how frequently this mistake is made.

Hadrian Jeffs
Norwich, Norfolk, UK

? On the tube

I photographed what looked like a huge tube of cloud floating just below a uniform blanket above rural Oxfordshire, UK, at 7.30 am on 11 December 2007. Anyone know why it formed?

Shuvra Mahmud
UK

Despite resembling a cigar rather than being saucer-shaped, such clouds are formed by a variant of the same process as that described in the previous question.

Like the slightly more familiar circular lenticular clouds, cylindrical clouds are fashioned by rotor currents, when air rising through a region of high humidity abruptly descends once more and the water condenses out to form droplets. Usually this takes place along the lee side of mountain ranges or high ridges, such as those found in the Owen Stanley range in Papua New Guinea and along Germany's Rhine valley. Circular clouds form when the currents mould the water droplets like clay thrown onto a spinning potter's wheel, while a cloud cigar such as this is shaped like a length of dough being rolled on a pastry board.

Although cloud cigars are most common over mountainous regions, they also appear in areas where there is no obvious topographical cause for rotor currents. They have been frequently seen over the Netherlands which, like rural Oxfordshire, is not an area noted for its mountainous terrain. So it seems possible that cloud cigars may also be created by microclimatic effects such as convection currents generated by heat from large urban or industrial areas.

The best places to see cloud cigars are the Massif Central in France and the Alps, the former being a hotspot for sightings during the great French UFO flap of the mid-1950s, and the latter the location for numerous reports of huge cylindrical 'foo fighters' by British fighter pilots during the Second World War. It is easy to imagine how the seemingly smooth surface might lead a witness to believe they are observing an artificial object. The giveaway is that almost all UFO reports of cloud cigars mention a vaporous exhaust, which is simply caused by wisps of cloud breaking from the main body.

Hadrian Jeffs
Norwich, UK

❓ Sky sports

How do clouds like this form?

B. J. Baxter
Ilford, Essex, UK

The photograph appears to have been digitally manipulated to enhance the effect. If it has not, the phenomenon depicted is exceptional in its intensity – Ed

This cloud form is known as mamma (the Latin word for breast or udder) and is technically described as a 'supplementary cloud form'. It is created when downdraughts bring cold air from higher levels, causing the air to reach its dew point and condense into cloud droplets. Compensating warm air rises between the individual pouches of falling air.

Mamma can form beneath various cloud types, including cirrus, cirrocumulus, altocumulus, altostratus and stratocumulus, where they often appear irregular in shape. However, beneath the overhanging 'anvils' of cumulonimbus, where heat has been lost to the atmosphere from the top of the anvil, they are often sharply defined pouches, as shown here. Mamma sometimes take the form of long contorted tubes that resemble the intertwined trunks of elephants.

Storm Dunlop
Chichester, West Sussex, UK

The pendulous features at the base of the cloud appear to be mamma (also known as mammatus or mammatocumulus), and are probably on the base of a cumulonimbus or storm cloud. Mamma occur when the upper parts of the cloud radiate heat into the atmosphere, cool and sink. If the sinking air is relatively warm and humid, the water vapour it contains will condense into cloud droplets as it mixes with colder, drier air beneath the cloud.

The process is an upside-down version of the way cumulus clouds form – the air associated with these warms at ground level and rises, its water vapour condensing to form clouds. Mamma air in the troposphere cools and sinks to form the clouds. Mamma attached to a cumulonimbus are associated with severe weather conditions, and aviators are strongly advised to avoid them.

A good summary of this phenomenon can be found in Gavin Pretor-Pinney's *The Cloudspotter's Guide* (Sceptre, 2006).

Ed Hutchinson
Cambridge, UK

? Cloud cover

A group of cumulus clouds was passing over my house in Tuscany, Italy, on a September afternoon lit by the low sun at about 5 pm. A clear shadow of the lower clouds appeared on the undersurface of the upper clouds (see below). Given that the light came from above both sets of clouds, how was the shadow projected on the higher ones?

Alessandro Saragosa
Terranuova, Italy

We received lots of theories for this phenomenon, but no single conclusive idea – Ed.

I took this photograph (opposite) in August at about 5 pm. The sun did not set for another 3 or 4 hours. Like your correspondent, I was mystified by the apparent shadow cast above, rather than below the cloud. This puzzlement was compounded by the rays of sunlight visible against the shadow.

Jeanette Stafford
Glasgow, UK

The original photograph sent by the questioner contains most of the clues to what is going on here.

There are three layers of cloud in the picture: an upper broken stratus and two lower cumulus layers. The sunlight, which seems to be coming from the lower middle of the picture, is falling onto all the cloud layers, including a small part of the lowest cumulus cloud. This can be seen by the short bright edge on this cloud. The light is strongly reflected and scattered, illuminating a large area of the underside of the middle layer, as the picture shows. This cloud, in turn, casts a shadow on the upper stratus cloud.

Chris Daniel
Kingston upon Thames, Surrey, UK

The sun is at a low angle, close to sunset, so it is illuminating the clouds from underneath. Shadows from the cumulus are being cast upward, resulting in this unusual display of light and shadow.

Rachel Vis
Melbourne, Australia

The shadow is not being projected onto the higher clouds, but rather onto the air between the cumulus clouds and the camera. I have seen similar shadows just before dusk caused by planes, which showed up as dark streaks. The scattered light makes the air appear to glow, except in the areas where it is being shaded by the clouds.

Bob Mitchell
Niceville, Florida, US

❓ Mountain headgear

I've seen mountain-top clouds similar to the one in this astounding image of a sombrero-like cloud atop Mount Fuji. What causes them? Do they only occur above volcanoes or do they occur above any mountain of suitable height?

Kevin Enright
Ironbridge, Shropshire, UK

This is a striking example of the cloud species known as alto-cumulus lenticularis (known in English as lenticular altocumulus – see also page 78). Such cloud forms when a stable, humid layer is forced to rise above the level at which condensation usually occurs, normally as part of wave motion. The uplift occurs above or downwind of an obstacle, and is certainly not restricted to volcanoes.

Depending on the exact atmospheric conditions, long trains of waves and clouds may be produced, and wave clouds have been observed far from any obvious source of motion. Such clouds tend to remain stationary as long as conditions, including wind strength and direction, remain constant. It is not uncommon for a series of humid layers to be affected, giving rise to a vertical stack of lenticular clouds separated by clear air, known to meteorologists as *pile d'assiettes* ('pile of plates').

Just as the air is forced to ascend and cool, producing condensation, so as the air descends at the rear of the wave it warms and any cloud dissipates. Close examination (with binoculars) will often reveal how the cloud is forming on the upwind side and dispersing at the trailing edge.

This cloud species is related to a similar cloud known as pileus, where an ephemeral cap of cloud forms as air is forced upwards above an actively rising cloud tower. In this case, the convective cloud cell often breaks through the pileus, producing a collar of cloud that is normally entrained into the rising column.

Storm Dunlop
Chichester, West Sussex, UK

This cloud type is a very well known phenomenon to glider pilots worldwide. The upper winds are blowing away from the camera towards Mount Fuji, probably at speeds in excess of 100 km/h at the mountain's peak. As the wind strikes the slopes of the mountain, it is forced to rise and becomes colder and less dense, and the moisture in the airstream condenses out, creating what is called a cap cloud over the peak.

Immediately downwind of the peak – the other side of the peak in the photo – the airstream, through a combination of temperature and stability characteristics, spills down the mountainside and then rebounds upwards again before descending even further downwind. With the right atmospheric conditions, this series of stationary vertical oscillations, or standing waves, can continue for 100 kilometres or more downwind.

The top of each standing wave is often marked by what is

called a lenticular cloud a few kilometres behind the mountain crest. The lenticular cloud appears stationary despite the very high wind speeds through it.

Massive standing-wave systems and their accompanying lenticular clouds are found above mountain systems all over the world, including Europe, the Andes, California and New Zealand, where the altitude record for a glider was set at 15,000 metres.

A miniature version of a standing wave can often be seen when a shallow flow of fast-moving water in a stream or gutter rides over a submerged object. This often creates two or three standing waves in the water downstream from the submerged object.

Max Hedt
Horsham, Victoria, Australia

What a beautiful lennie! As any glider pilot will tell you, this is a lenticular cloud formed in a standing wave. When flying a glider in such a wave, the ride is incredibly smooth, and if the wind speed is higher than the glider's stall speed then you can head directly into wind and seemingly be suspended in space with just the aircraft instruments indicating that the glider is actually flying. Magic!

Mike Debney
Melton South, Victoria, Australia

❓ Genius of the lamps

On a September evening in 2003, my husband took this photograph (see below) outside our house in Cambridge looking east. The clouds seem to have come out of Aladdin's lamp. Can someone explain how such a remarkable pattern can be formed, or is the genie's tale true?

Omara Williams
Cambridge, UK

Within a few days of the original photograph of the 'Aladdin's clouds' being published, we received three more photographs taken of the same five clouds on the same day from different locations in eastern England (see the following two pages). One of them [A] was taken by Rowan Moore at Dunchurch, near Rugby, 100 kilometres to the west of Cambridge; one [B] was taken by Martin Williams at Holme, near Peterborough, 40 kilometres north-west of Cambridge; and the third photograph [C] by Clive Semmens in Ely, 25 kilometres north of Cambridge. The final picture [D] is of a larger group of similarly shaped clouds photographed from Nottingham, in central England, by Sean May on a different date. The explanation for these peculiar clouds has been provided by another reader... – Ed.

A

B

These strikingly shaped clouds are not rising and trailing particles below them, as their appearance in the photographs suggest. They are actually altocumulus clouds from which precipitation is falling.

Such trails of water droplets or ice particles are called virgae (or sometimes fallstreaks) and, by definition, do not reach the ground. Precipitation that does reach the Earth's surface is known technically as praecipitatio.

Virgae are produced from what are known as heads. These can be distinct, rounded clumps of cloud like the ones that are

C

D

shown here, ragged tufts, or extremely small patches of cloud that are difficult to distinguish from the tops of the virgae themselves. They occur at all cloud levels, and ordinary cirrus clouds – the thin wispy clouds that are popularly called mares' tails – essentially consist solely of virgae.

Virgae often show distinct bends like the ones clearly visible in the photographs. These bends often occur where the falling ice particles melt into water droplets.

The ice crystals fall almost vertically. But as the water droplets start to evaporate, they become smaller and so fall more slowly, leaving them trailing behind the cloud above.

In other cases, the bend may indicate a region of wind shear, where the strength or direction of the wind changes. On very rare occasions, where the wind is stronger at a lower level, virgae have even been observed in front of the head that generated them.

Storm Dunlop
Chichester, West Sussex, UK

❓ Clear flight path

This photograph was taken near Maldon, Essex, in the UK, looking directly overhead. It appears to show the result of an aircraft flying through thin cloud and dispersing it along its flight path. If an aircraft was responsible it had long since passed when the picture was taken. Is this a common sight and what mix of conditions is required to produce the effect?

Neil Sinclair
Chelmsford, Essex, UK

This is a relatively common occurrence, known as a dissipation trail or distrail. Depending on the exact circumstances, one of three mechanisms may be involved. First, the heat from the aircraft's engines may be sufficient to evaporate the cloud droplets. Second, the wake vortices shed by the wings may mix drier air into the cloud, lowering the relative humidity and again causing droplets to evaporate. Finally, the exhaust may introduce glaciation nuclei into the cloud. These are particles around which ice crystals form, causing freezing to occur. The crystals then fall out of the cloud. This is a very common mechanism, but the photograph shows no sign of falling trails of ice, which are known as virgae.

The first mechanism seems to be rare and is not accepted by some authorities, so the vortex explanation is probably the most likely in this case.

Storm Dunlop
Chichester, West Sussex, UK

❓ Misty morn

*While camping in the desert north of Coober Pedy, South Australia,
in July 2007, my son and I were privileged to witness a white
rainbow at daybreak, similar to the one shown here. The landscape
was covered in mist and the white rainbow's arc seemed to grow as
the sun came up, though it faded away as the mist evaporated. I've
asked old swagmen, indigenous locals and several lecturers at two
universities, but no one has ever witnessed the phenomenon. What
caused it?*

Fred Richardson
Alice Springs, Northern Territory, Australia

This was a fogbow, sometimes known as a cloudbow or mistbow.
Like a primary rainbow, it is centred on the point opposite the
sun and has an angular radius of approximately 42 degrees. It is
caused by the same mechanism: reflection and refraction of sunlight by water droplets. In this case the droplets are unusually
small – less than 50 micrometres across – allowing diffraction

to spread the bands of colours so that they overlap and appear white.

Occasionally, fogbows will show a bluish tinge to the inner edge and a reddish one on the outer. From some viewpoints, such as an aircraft, a fogbow can appear as an almost complete circle.

Storm Dunlop
Chichester, West Sussex, UK

A fogbow is a frustrated rainbow, formed in essentially the same way. Sunlight destined to create rainbows and fogbows is refracted twice – once as it enters a water drop and again as it leaves. While inside the water drop between the two refractions, the light bounces off the inside back surface, sending it heading back towards the sun. This is why rainbows and fogbows are seen when the sun is behind the observer. The path of blue light is bent more than red by the droplet, usually causing sunlight to be dispersed into the colours of the visible spectrum with blue at the bottom of a rainbow and red at the top.

Rainbows appear white when water droplets are less than 100 micrometres across – small enough for diffraction to dominate over refraction. Each water drop forms its own diffraction pattern – bands of alternate constructive and destructive interference – for each colour: the smaller the drop, the broader the bands. When the drops are small enough, these bands become so broad that all the colours overlap, essentially mixing them all together again to make white.

Beautiful images can be found at www.atoptics.co.uk/droplets/fogbow.htm.

Mike Follows
Willenhall, West Midlands, UK

❓ Hair-raising event

Walking along the breakwater at Berwick-upon-Tweed in north-east England, my granddaughter and her mother noticed their hair was standing on end. It started to rain soon afterwards, but there was no thunder or lightning that day. What was happening?

Richard Turner
Harrogate, North Yorkshire, UK

We have answered this question before in an earlier book, but this time we are able to use the startling photograph the questioner supplied originally – Ed.

From one of my physics textbooks I recall a hair-raising picture of a woman standing on an exposed viewing platform at Sequoia National Park in California. She was in grave danger. Lightning struck only minutes after she left, killing one person and injuring seven others (*Fundamentals of Physics*, 6th Edition, by David Halliday, Robert Resnick and Jearl Walker, published by John Wiley and Sons). It's likely that similar conditions were abroad on the day your photo was taken.

Most lightning clouds carry a negative charge at their base. Anything close to the cloud would feel the effect of electrostatic forces: electrons in a person's hair would be repelled downwards, leaving the ends of the hair positively charged. The positive hair tips are then attracted to the cloud – and repelled by each other – and stand on end. It's rather like rubbing a balloon on someone's hair to make the hair stand on end: the balloon becomes negatively charged and the hair is attracted to it.

Lightning victims often describe how they felt tingly and their hair stood on end before they were struck. Fortunately air is a good electrical insulator and, in this instance, the charge in the clouds wasn't high enough to jump down to earth, so there was no lightning. However, this was probably a lucky escape for your family. If your hair stands on end outdoors or your skin is tingling, lightning may be imminent and it's best to run for suitable shelter.

Iain Longstaff
Linlithgow, West Lothian, UK

The phenomenon described above is known as luck – the two people were fortunate not to have been struck by lightning. Experienced hikers and climbers know that this hair-raising phenomenon can be a precursor to a lightning strike and are taught to flatten themselves or, if climbing, dive for lower ground.

There is a vertical voltage gradient in the atmosphere,

typically in the order of 100 volts per metre on a clear, dry day. For an average adult male, then, there will be a 180 to 200-volt difference between the toes and the top of their head.

When electrically charged would-be storm clouds scud overhead, an induced ground charge follows the clouds, markedly increasing the voltage gradient. If the potential difference is sufficient to overcome the resistance of the air – around 3 million volts per metre – then lightning leaps across the gap. In practice, lightning strikes are possible at substantially lower voltage differences. The fact that the reader's family saw no lightning and heard no thunder merely suggests that, luckily, the voltage never rose high enough for a lightning strike.

Larry Constantine
Department of Mathematics and Engineering
University of Madeira
Funchal, Portugal

❓ Earn those stripes

Walking on the beach in Waiheke Island, New Zealand, I spotted an interesting cloud formation consisting of small stripes. As I watched, more followed. The stripes gradually got longer and longer, and then slowly disappeared. How did these strange formations occur?

Brendan Zwaan
Waiheke Island, New Zealand

Stripes of clouds in repetitive patterns are ripples in the surface of a layer of humid air beneath a layer moving in another direction, much like ripples on calm water beneath a breeze. As air passes through such a region, it warms on the way down and cools as it rises. Usually we cannot see this happening, but if the humidity is such that the cooling air condenses the moisture, we see the cloud as a streak at the crest of the ripple. Air passing through the ripple crest doesn't stop, it moves down and warms up, evaporating the cloud and then re-forming it in the next crest.

So those apparently stable streaks of cloud are an illusion. They are simply cool zones through which air ripples. Even as you watch, the droplets in any streak vanish, continuously and invisibly being replaced by incoming water droplets. Their vapour flows on to the next peak where it re-condenses and passes through the next streak in turn. Just as ripples on a pond can change, so the length and interference patterns of cloud ripples may change, sometimes bewilderingly.

Jon Richfield
Somerset West, South Africa

The photograph shows billow clouds created by wind shear. It is a manifestation of a kind of turbulence known as Kelvin-Helmholtz instability, which is created when different layers of air move at different speeds. Typically, the upper layer moves faster than the lower layer. This creates eddy currents or an oscillation at the boundary between the layers, creating a ripple effect.

Airline pilots normally take a detour around billow clouds because they betray the presence of potentially dangerous turbulence.

Mike Follows
Willenhall, West Midlands, UK

4 In your kitchen

❓ Bad soap

I found this forgotten bar of soap after winter at my home in northern Sardinia. It had grown a coat of mould. What is the mould and how did it grow on soap, which is supposed to keep your hands clean?

Patrizia Figoli Turchetti
Bellaire, Texas, US

We use soap for cleaning because it is a detergent – a means of emulsifying insoluble, largely fatty, dirt in water. Its nutritional value is usually irrelevant, but pure traditional soap consists of fatty-acid salts. Because of this, it is completely digestible in modest quantities. You may see a dog scoffing a chunk of soap because it smells appetisingly of fatty acids, but only if it doesn't contain too much scent or lye (sodium hydroxide), which is used in the production process. Missionaries who introduced soap to some tribal communities in Africa were startled to find that members of their congregations would treasure a fatty-tasting bar as a treat, occasionally licking a finger that had been moistened and rubbed on the soap.

Toilet soap commonly contains surprising amounts of starches, oils, glycerol and other materials that make it smoother, less aggressive to the skin or simply cheaper to produce. These are all edible too, and moulds are happy to consume them. As long as the soap doesn't contain too much sodium and the air is moist enough, as it might well be in a bathroom, a bar of soap can certainly grow some very contented fungi.

I suspect your soap sported a selection of Sardinian domestic moulds. *Fusarium*, *Mucor*, even white strains of cheese fungi such as *Penicillium camemberti* might be present. They are probably harmless. Try some if you like…

Jon Richfield
Somerset West, South Africa

Though we admire Jon Richfield's enthusiasm for direct scientific experimentation, we recommend that you do not eat the soap before finding out exactly what is growing on it – Ed.

While I can't help your reader identify the strain of mould on the soap, I can explain how mould can grow on something that is used for cleaning your hands.

Soap consists of long-chain organic molecules, with one end that is polar (charged) and the other non-polar (uncharged). The polar end readily dissolves in water, which is also polar, while the non-polar chain readily attaches to grease and oil, which are similarly non-polar.

The soap therefore acts as a go-between: one end attaches to the oil and the other end wants to be in the water. This enables the oil on your skin to dissolve and be washed off. Without the soap, the polar water molecules would rather stick together than attach themselves to the oil on your hands.

Mould can grow on all sorts of apparently uninviting materials, including leather and wallpaper. Soap is no different – it's just another organic material. This is not really a paradox. After all, when you use soap you don't actually leave it on your hands in order for them to stay clean. You wash it off.

Simon Iveson
School of Engineering
University of Newcastle
New South Wales, Australia

? Ripple effect

The glass in these photos seems to have no ripples in it when viewed from the side, but lots when viewed from above. Why is this?

Liam (aged 11)
Galway, Ireland

The 'ripples' are caused by multiple reflections as light from the base of the glass bounces between the inner and outer surface of the glass on its way up. At each reflection a little light escapes and enters the eye, producing the observed 'ring' at that height. If you look down the outside of the glass from above, you will see ripples once again.

Keith Thompson
Bendigo, Victoria, Australia

The effect is caused by internal reflection, just as happens inside an optical fibre. To see a neat demonstration of this, cover the sides and one end of a glass tube with black plastic tape, then punch a hole about 1 millimetre across in the sealed end. Now look down the small hole and you will see concentric rings caused by internal reflections. If you look through the unsealed end you get some pretty interesting views too.

I have used this effect as a little test when interviewing prospective physics students at Oxford.

Mike Glazer
Clarendon Laboratory
University of Oxford, UK

❓ Watermelon weirdness

What caused the oddly attractive patterns on the skin of this watermelon? It was grown and purchased on one of the Canary Islands.

Peter Drake
By email; no postal address supplied

The patterns are caused by papaya ringspot virus type-W, which despite its name only infects plants such as watermelons, cucumbers and courgettes.

While it can stunt growth and reduce yield, the patterns on the skin are, as you say, attractive, and in some parts of the world such fruit sell for higher prices. The virus is classified as a potyvirus and like most plant viruses has a single-stranded RNA genome.

Meriel Jones
Port Sunlight, Merseyside, UK

❓ Thinking person's crumpet

A little while ago we froze a packet of those pancake-like products that in these parts we call crumpets. At the time, the sealed plastic packet seemed to contain a lot of air, but after four months in our freezer it had contracted tightly against the crumpets, which had also shrunk. After two more months at room temperature, the whole package appeared to have shrunk still further, though with no sign of mould or decay. The ingredients are listed as flour, water, yeast, raising agents, E450, E500, salt, sugar, preservative, calcium propionate. What's going on?

Chris Greenwood
Ettington, Warwickshire, UK

Our thanks to Warburtons, the company that made these crumpets, for the following – Ed.

Warburtons crumpets are a short shelf-life, high-moisture product and as such, are particularly susceptible to food spoilage organisms. The product is packaged in a carbon-dioxide environment to extend the shelf life and to protect the crumpets against microbial spoilage, particularly aerobic organisms.

Over time, the carbon dioxide gas is absorbed by the liquid in the product, and as it takes up less space in liquid form this reduces the internal pressure of the package. As long as the seals are intact, the differential in pressure means the packaging contracts. The lack of spoilage in the product is an indication that the seals are intact and the integrity of the product remains unaffected.

Claire Minzey
Clarion Communications
London, UK

❓ No use crying

While making a cup of coffee I spilled some milk, and it made an interesting pattern (see below). There were approximately 18 small droplets surrounding a larger central droplet. It reminded me of a photograph I saw in a textbook during my childhood, where a drop had just fallen into a glass of milk, resulting in a splash like a king's crown. Why did my pattern and the one from the book form? Presumably they are related. Do other liquids make similar patterns?

Stephen Broderick
Toowoomba, Queensland, Australia

The pattern of droplets surrounding the larger drop of spilt milk is the result of corona splashing. The study of splashes has a long and distinguished history, dating back to the late 19th century. Arthur Worthington, who was one of the first people to systematically study drop impacts, made it his life's work, and his book *A Study of Splashes* (Longman, 1908) contains many fascinating photographs of splashes. However, perhaps the best-known photographs of a corona splash are those captured by Harold Edgerton starting in 1936.

When a drop hits a surface, it can either spread, splash or rebound. The outcome depends on a number of factors, including impact speed, size, density, viscosity and whether the surface is initially dry or damp. For dry surfaces, impact drop morphologies are also influenced by the roughness and 'wetability' of the surface.

On a typical dry, flat and relatively smooth bench, such as the type often found in kitchens, the moment the falling drop strikes the surface, portions of the liquid moving downwards are pushed outwards from beneath the collapsing drop and immediately begin to spread along the bench top. A narrow and highly curved neck therefore develops between the thin liquid layer and the largely uncollapsed spherical drop above it. Provided the impact speed is above a critical value and the surface is not too smooth, surface tension in the necked region gives rise to a force directed at an angle to the horizontal.

In corona splashing, the upward component in the surface tension at the neck causes the leading edge of the radially spreading liquid film to bend upwards, giving rise to the formation of a 'corona', or crown, of the type mentioned in the question. Excess pressure caused by the collapsing drop continues to push liquid into the walls of the crown as the impact unfolds, extending it outwards and upwards.

As the crown rises and increases in size, the rim slows and thickens, taking on a roughly toroidal shape. The underlying mechanism leading to the rim of the crown breaking up into many smaller droplets is not yet entirely understood. It is thought, however, that at the moment when liquid begins to feed into the walls of the crown, rough anomalies on the bench's surface initiate small ripples which rapidly grow and inevitably cause the rim of the crown to form cusps, which break up into thin jets directed upwards and away from the crown.

Instabilities in the ejected jets from the rim of the crown cause a droplet to pinch off at each end, giving rise to the distinctive coronet (as shown in the photo below) that is responsible for the observed splash pattern.

Shortly afterwards, the walls of the crown start to thicken as it slows before falling back onto the bench under gravity, together with what remains of the protruding jets, to form the larger central drop. The effects of surface tension are responsible for holding the collapsing coronet together during its final denouement.

Besides milk, corona splashing is found in many other liquids including water, various alcohols and some paints. However, the opaqueness of milk makes it easier to see the droplet ring resulting from a corona splash.

Seán Stewart
The Petroleum Institute
Abu Dhabi, United Arab Emirates

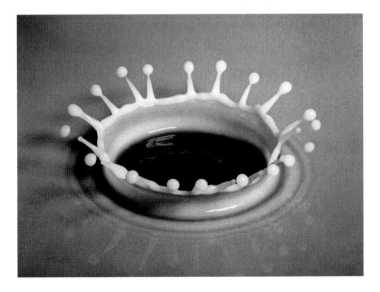

? Twin chicks

When I cracked open my breakfast boiled egg, I found a whole new egg inside. It was not a double-yolked egg, it was a double-egged egg – a completely new egg with a shell and yolk inside another (see below). Can anybody explain it?

Liam Spencer
York, UK

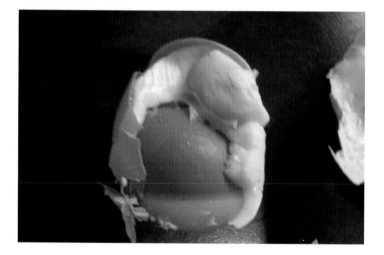

An egg within an egg is a very unusual occurrence. Normally, the production of a bird's egg starts with the release of the ovum from the ovary. It then travels down the oviduct, being wrapped in yolk, then albumen, then membranes, before it is finally encased in the shell and laid.

Occasionally an egg travels back up the oviduct, meets another egg travelling down it, and then becomes encased inside the second egg during the shell-adding process, thus creating an egg within an egg. Nobody knows for sure what causes the first egg to turn back, although one theory is that a sudden shock could cause this. Eggs within eggs have been reported in hens, guinea fowl, ducks and even Coturnix quail.

Incidentally, it is especially unusual to encounter this phenomenon in a shop-bought egg, because these are routinely candled (a bright light is held up to them to examine the contents), and any irregularities are normally rejected.

Alex Williams
Haverfordwest, Pembrokeshire, UK

As the curator of the British Natural History Museum egg collection, I've come across quite a few examples of egg oddities. Double eggs (as opposed to multiple-yolked eggs) are less common than some other oological anomalies and consequently the 'ovum in ovo', as the phenomenon described here is known, has attracted specific scholarly attention for hundreds of years.

The Dominican friar and polymath Albertus Magnus mentioned an 'egg with two shells' as far back as 1250 in his book *De Animalibus*. By the late 17th century, pioneering anatomists like William Harvey, Claude Perrault and Johann Sigismund Elsholtz had also given the phenomenon their attention.

Four general types occur – variations of yolkless and complete eggs – but this form in which a complete egg is found within a complete egg is relatively rare. Several theories have been proposed for the origin of double eggs, but the most likely suggests that the normal rhythmic muscular action, or peristalsis, that moves a developing egg down the oviduct malfunctions in some way.

A series of abnormal contractions could force a complete or semi-complete egg back up the oviduct, and should this egg meet another developing egg travelling normally down the oviduct, the latter can engulf the former; more simply, another layer of albumen and shell can form around the original egg.

Often when no yolk is found within the 'dwarf' or interior egg a foreign object is found in its centre. This object has served as a nucleus around which the albumen and shell were laid down, in a process not dissimilar to the creation of a pearl.

The Delaware Museum of Natural History in the US has a fantastic example in its collections (see photo below).

Anybody interested in learning more about this subject should try to find a copy of *The Avian Egg* by Alexis Romanoff and Anastasia Romanoff (John Wiley & Sons, 1949) and turn to pages 286 to 295.

Douglas Russell
Curator, bird group
Department of Zoology
The Natural History Museum
Tring, Hertfordshire, UK

❓ It's no yolk

At breakfast one day, my 4-year-old son was peeling his boiled egg when he noticed something unusual. On the side of the egg was an almost perfect white square with sides about 1 centimetre long. This was enclosed in an oval patch of yellow matter. Although the yellow patch looked like egg yolk, the egg appeared normal when I cut it in half, and I could not see any connection between the egg yolk in the centre and the yellow patch on the outside. Does anyone know how this formation was created?

Johan Forsberg
Linköping, Sweden

The answer from Paul Hudgins was favoured by several other people.
But if this is the explanation, why isn't this pattern seen more often,
and what caused the mysterious yellow colouring? The case remains
open – Ed.

Some eggs have a very soft shell when laid. The imprint was likely caused by the pressure of the hen sitting on the egg and pressing it against the wire mesh forming the bottom of its cage. This would have damaged the membrane that surrounds the egg, producing the square pattern.

Paul Hudgins
Jacksonville, Florida, US

One can tell from the picture that at one time the egg had been standing on a surface where a square support had been in contact with the shell, but where a surrounding pool or pad had applied a fluid that had diffused through. Whether the invading substance was itself yellow or whether it just changed the colour of the albumen is hard to say. We can, however, be sure that no internal influence could have created such a precise and shallow pattern.

The architecture of an eggshell is amazing. It protects the contents from outside threats, whether microbial, chemical or mechanical. It permits carbon dioxide to escape and lets oxygen in. It supplies calcium for the chick's skeleton. Eventually it lets the hatchling out.

The protection is not perfect, however. By writing on the shell with suitable inks, or using vinegar or alum, one can cause marks that become visible on the hard-boiled egg white as the dyes seep through. The picture clearly shows an accidental example of some such effect. It's possible, for instance, that some fluid had spilled onto the scale that grades eggs by weight.

Jon Richfield
Somerset West, South Africa

❓ Green eggs and cabbage

While cooking recently, I found out quite by accident that if you pour the juice from cooked red cabbage over a frying egg, the egg white turns green. Why is this?

Graham Cooke
Dublin, Ireland

I teach a Year 7 class of 11- and 12-year-olds at High Storrs School, Sheffield. The answer below is the result of experimentation by the class.

We thought we knew the answer, but we wanted to be sure. We knew that red-cabbage juice (which is actually purple) is a good indicator of whether a substance is an acid or an alkali. We then attempted to change the juice's colour with citric acid, which turned it red, and with ammonia, which changed it to green because it is an alkali. When we added the cabbage juice to raw egg it turned green, so we established that egg white, or albumen, is alkaline. We checked this with universal indicator and discovered that the pH of egg white is about 9, meaning it is alkaline, as we suspected. Cooking the egg after this made no difference to its colour.

Philip Ward and Class 7C
High Storrs School
Sheffield, UK

There is a well-known high-school science experiment that involves creating a 'red-cabbage indicator' to measure the pH of acids and bases. Red-cabbage juice is added to solutions of known pH and a chart of the colour change observed for each pH is created.

The juice of a red cabbage contains a water-soluble pigment molecule called flavin, an anthocyanin, which is also present in plums, apple skins and grapes. Adding an acid (which has a pH value of less than 7) to the indicator will change this pigment to red, while a base, or alkali (which has a pH greater than 7), will change it to a green/yellow colour. In neutral solutions (pH 7) it is purple.

The colour changes in the juice are the result of a change in its hydrogen ion concentration after an acid or base is mixed with it. In an aqueous solution, acids donate hydrogen ions and bases receive them.

The colour change seen in the egg white shows that egg whites are basic and have a pH of approximately 10.

Emma Bland
East Malvern, Victoria, Australia

Anthocyanins are versatile pH indicators: they change colour depending on the hydrogen ion (proton) concentration or pH around them. They have a three-ringed structure with light-absorbing properties that vary according to how many protons are attached. In acidic conditions, the molecule acquires protons and turns red or pink.

'Red' cabbage is only red after pickling in acidic vinegar. If conditions are neutral it is purple and if alkaline it loses protons, turning blue to green to yellow, depending on the strength of the alkali. The questioner has shown that the pH of egg white is alkaline, probably about pH 10.

The colour changes are quite subtle and match those of many commercially produced indicators. The only unfortunate side effects are the smell cabbage creates and the fact that the indicator goes off quickly. Nonetheless, you can have great fun extracting the indicator (boil shredded purple cabbage in a minimum of water and strain off the indicator liquid when cool) and using it to determine the pH of substances and conducting neutralisation experiments. Obviously, if you try this on strong solutions be very careful.

All this means hours of fun when you should be cooking, which is why I am now barred from buying the stuff.

Kate Johnston
The Centre for Plant Sciences
University of Leeds, UK

You can create 'magic' paper to amaze children by soaking reasonably low-quality paper in boiled red-cabbage water and leaving it to dry. Let the children paint on it with household substances such as vinegar and washing powder dissolved in water and they'll see a range of colours, including pink, blue, yellow and green, depending on the pH and strength of the substance.

Susan Crafer,
Barnsley, South Yorkshire, UK

Anthocyanins are powerful antioxidants that are thought to protect the plant's photosynthesising apparatus against free radicals produced under bright light, especially in cold weather. The red cabbage, therefore, is a green cabbage with a shield of anthocyanins above its chloroplasts. This shield absorbs the photo-oxidative and unusable wavelengths of green light, and cancels out the green of chlorophyll.

Boiling up the cabbage, like boiling any leaf vegetable, ruptures the leaf cells, allowing the chlorophyll and anthocyanin pigments to leach into the water. This becomes a mixture of green from the chlorophyll and purple-blue from the anthocyanins.

Caspar Chater
Centre for Economic Botany
Royal Botanic Gardens, Kew
Surrey, UK

The photo on the previous page illustrates the phenomenon well. It was supplied by Peter Scott of Hove, East Sussex, UK – Ed.

❓ Funny onion

If, after peeling an onion then topping and tailing it, I wait before chopping it, the middle segments push up and outwards. Why?

Alan Middleton
Weymouth, Dorset, UK

An onion is a bulb, with each layer being a specialised leaf. In your vegetable rack, it was simply biding its time. If you had put it in damp soil, the innermost leaves would have grown, emerging from the top of the onion and turning into the green parts of a new onion plant. Your kitchen-table surgery has enabled the first stages of this growth process to start.

For any plant cell to grow it needs three things: an external supply of water; an internal supply of solutes, such as sugars and salts; and – most crucially in this case – a lack of mechanical constraint hemming the cell in.

Even in your vegetable rack, the inner leaf cells find a supply of water. Water is constantly entering and leaking from any given cell, and for the inner leaves' cells to grow they need to take in some of the water escaping from the cells of the outer leaves.

As the cell sucks in water by osmosis, an internal pressure called turgor builds up, which tends to swell the cell, driving growth. In a fresh, juicy onion, all the cells have turgor but most of them can't grow because their cell walls are too tough.

On the other hand, the bulb's innermost leaves can potentially grow, because they have thin cell walls, easily stretched by turgor. What stops them is the tough outer leaves of the bulb, which act as a corset, mechanically constraining the inner leaves. When you slice the bottom off the bulb, the base of the 'corset' is ruptured and the inner leaves can now grow lengthwise, so they protrude as seen in the photo.

Incidentally, the onion in the photograph is upside-down – its inner leaves are protruding from the bottom end of the bulb. In damp soil, the leaves would emerge from the top end, growing up into the sunlight.

In this case, however, your correspondent has cut the bulb in such a way that the corset is still more or less intact at the top end and protrusion is possible only at the bottom end.

Stephen Fry
The Edinburgh Cell Wall Group
Institute of Molecular Plant Sciences
University of Edinburgh, UK

5 Goo and gardening

❓ Fly catcher

The fungus in the photo appeared on rotting, chopped-up trees used as mulch in our garden. It is about 10 centimetres in diameter and appears to feed on insects, and so has an appropriate stink of rotting flesh. What is it?

Ted Webber
Buderim, Queensland, Australia

The fungus is *Aseroe rubra*, literally 'disgusting red', whose common names include sea anemone, starfish and stinkhorn fungus. Found in Tasmania and as far north as south-east Queensland, it has the distinction of being the first fungus from Australia to be scientifically described. It was named, no doubt after he smelled it, by French botanist Jacques Labillardière at Recherche Bay, Tasmania, in 1792.

The fungus appears to do a good job of mimicking a wound on an animal – an interesting piece of evolution, if true. It looks a bit like a fly trap, but it doesn't catch or eat flies. Instead, it uses them to spread its spores. For food it uses wood in the mulch or forest litter that it inhabits.

The fruiting body in the photo shows the smelly black spore slime that acts as a fly attractant. I find its odour more like that of rotting teeth than rotting flesh, a concentrated *essence de caries*.

The fruiting body lasts only a few days but rewards quiet observation. Our *Aseroe* specimens attract the rather beautiful, though agriculturally troublesome, green blowfly.

Kevin Maher
Witta, Queensland, Australia

Curiously, this fungus has leapt several continents and established itself in the UK on Oxshott Heath in Surrey, where it has been appearing exotically for the past 10 years or more.

The fungal fruiting body doesn't catch flies any more than a ripe apple catches wasps. The flies (which tend to be mainly bluebottles in the UK) are attracted to the carrion smell of the slimy spore mass supported on the tentacle-like arms. The slime contains sugars and the flies ingest it, spores and all. These pass through the gut unharmed and are dispersed elsewhere. The fungus itself – visible as a white mycelium – lives on rotten, often buried, wood.

A related Australian species with longer tentacles, *Clathrus archeri* or the devil's fingers, has also become established in southern England and is now quite widespread, as is the European *Clathrus ruber*, or cage fungus, whose tentacles mesh to form a cage-like receptacle.

More common and more familiar is our native stinkhorn, which also lures flies to help spread its spores. The stinkhorn forms fruiting bodies as obvious as its Latin name, *Phallus impudicus*, suggests. It is said that Charles Darwin's daughter Etty used to rise early to destroy any that she found, in order to ensure that the morals of her maids were not corrupted.

Peter Roberts
Royal Botanic Gardens, Kew
Surrey, UK

? Shell suit

One of my solar-powered garden lights has become home to a colony of snails. Typically about 20 congregrate there every day, mainly on the north face of the light, but none on the solar panel. Are they adapting to 21st-century technology and using solar power to extend their periods of access to warmth and light?

Roger Bloor
Newcastle-under-Lyme, Staffordshire, UK

The snail pictured is the garden snail, *Helix aspersa*, which was accidentally introduced into England, probably in Roman times, by trade with mainland Europe. Among other habitats, it is associated with disturbed areas such as gardens and waste ground.

Avoiding loss of water is a major priority in the life of a snail, and is perhaps the single most important factor influencing their day-to-day activity and behaviour. As a result, snails are essentially reclusive animals, spending much of their lives hidden away in sheltering microhabitats. Gardeners will be well aware that they can be found in abundance under bricks or stones.

Garden snails commonly come together in places that offer some type of protection, in particular to undergo aestivation and hibernation. Aestivation is a period of inactivity in dry weather during which snails avoid losing moisture by withdrawing into their shell and adhering to a substrate via a mucous membrane. Hibernation occurs in winter and involves the shell being sealed with a calcareous plate, or epiphragm, to avoid water loss, while the snail's pulse rate is reduced.

In the situation your questioner describes, by choosing the north side of the solar-powered light, the snails both avoid exposure to the sun and are able to shelter in the artificial 'crevice' afforded by the lip of the solar panel. They would naturally avoid the top surface of the panel, which is much more exposed and, of course, faces the heat of the sun.

During warmer months the snails will often be active away from the lamp, returning to their 'roost' by following the chemical signals in their slime trails. At night, it is possible that the light actually provides them with some warmth. However, snails would normally avoid the desiccating heat and light given out by a more powerful lamp.

Peter Topley
Bedfordshire, UK

Snails' preference for smooth, sheltered surfaces forms the basis for a good snail trap requiring no poison. Where snails are most plentiful, stand plastic flowerpots upside down with their rims raised about 25 millimetres off the ground on the side facing away from the sun. After a spell of rain following dry weather, you can collect these garden guests by the dozen for donation to grateful hedgehogs and neighbours.

Jon Richfield
Somerset West, South Africa

Check that your neighbours are happy with this arrangement before donating your snails – Ed.

? Slime time

I travelled to Scotland recently where I took this picture of an organism. It was smooth, glossy and transparent like melting ice, and quite hard to the touch. It was not alone; there were four others close by, the biggest of which was about 20 centimetres across. Can any reader tell me what it is?

Alexander Markov
London, UK

There are lots of ideas about what this object is, none of which is conclusive – Ed.

The organism is most likely a fungus from the family Phanero-chaetaceae called *Phlebiopsis gigantea*. As with many organisms, the naming of this fungus has troubled taxonomists, so in some books it is called *Phlebia gigantea* and in others *Peniophora gigantea*. The fungus is interesting because it has been deliberately introduced by humans into pine forests as a biological control against the timber-spoiling fungus *Heterobasidion annosum* from the family Bondarzewiaceae.

A less likely candidate is a fungus from the related family Cyphellaceae called *Radulomyces confluens*. The description of the organism as being glossy, almost transparent and quite hard to the touch rules out slime mould (Myxomycetes), contrary to what the question's headline suggests.

Peter Cook
Withernsea, East Yorkshire, UK

I have two thoughts on this. The first is that it is 'star jelly' (also known as 'witches' butter'), more correctly described as the cyanobacterium nostoc, which forms jelly-like masses. The second is that it is frozen human excrement dropped from a toilet of a passing plane.

Rich Boden
Department of Biological Sciences
University of Warwick
Coventry, UK

This might be unfertilised frog or toad spawn. The spawn is formed prior to breeding and is stored in the amphibian's body before it is fertilised. When exposed to air it spontaneously explodes into slime. This substance can be seen in late winter or early spring, and is attributed to predation by herons. These birds can eat large numbers of frogs or toads, conveniently eating the body but not the spawn – leaving no clue as to where the slime came from.

Colin Campbell
Science leader
Soils Group, Macaulay Institute
Aberdeen, UK

A lack of information about the habitat of the organisms makes it hard to be accurate, but it looks as if it could be the seashore, in which case what the reader observed could be a group of simple tunicates, also known as sea squirts, urochordates or ascidians. These are marine animals which begin life as tiny tadpole-like creatures that settle down on rocks or seaweed to become adults. They then become covered in a coat of cellulose and spend their life filtering food from seawater. They are occasionally washed up on the beach at Portobello near my home after rough seas.

If the object is hard, it's unlikely to be the most common species, *Ciona intestinalis*, so it could be *Ascidiella aspersa*. A larger species is *Phallusia mammillata*, which is covered with rounded elevations, hence its specific name.

Tim Bolton-Maggs
Edinburgh, UK

The organism is the pupa of a haggis. Your correspondent is very fortunate: haggis normally pupate only under cover of dense heather or bracken, and pupation usually happens at the onset of the Scottish winter in early September. In due course the outer skin becomes opaque, and the inner tissues become firm and granular as water is lost. It is at this point that pupating haggis are collected for human consumption. Those that elude would-be gourmets remain in a dormant state until spring, emerging as fully developed adults or 'imagos', ready to seek a mate and ensure the continued supply of this delicacy.

Maxwell Buchanan
Cambridge, UK

❓ Leaf antlers

What causes these strange horny growths on tree leaves?

Mike Child
Bedford, UK

These 'antlers' are almost certainly galls. Many plants, both herbaceous and woody, react to irritants by producing a tumour-like growth around that irritant. The irritant may be a virus or bacterium, a fungus, another plant, a nematode worm or one of several groups of arthropods – insects and mites – which suck the sap or lay eggs on or in the plant. The larvae of insects that cause galls live in its centre, eating away happily. They will usually pupate when the gall falls to the ground into the autumn leaf litter and the adult insect emerges the following spring, in time to attack the following year's spring growth.

Usually, the gall has a characteristic form related both to the plant species and to the irritant species. The same irritant may induce different shaped galls in different plant species and also at different stages of its life cycle. The excellent *British Plant Galls* by Margaret Redfern and Peter Shirley, published by the Field Studies Council (ISBN 9781851532148), illustrates many hundreds of galls.

So which gall is in the photo? There is no scale on the picture but the leaf is from a tree. It appears to be roughly egg-shaped with an irregularly and coarsely toothed margin. These features suggest the leaf is possibly from a poplar or lime tree. If that is so then the galls may be about 5 to 10 millimetres tall. *British Plant Galls* lists no galls of this general shape for poplars, but there are two likely candidates under limes, both are mite-induced. In the broad-leafed lime (*Tilia platyphyllos*) the mite is *Eriophyes tiliae*, producing a gall greater than 8 millimetres tall; in the small-leafed lime (*Tilia cordata*) the irritant is the mite *Aceria lateannulatus*, whose gall is about 5 millimetres tall. Both galls may also be found on the common or European lime (*Tilia* × *europaea*).

Mike Snow
Crymych, Pembrokeshire, UK

These are lime nail galls caused by the lime nail gall mite (*Eriophyes tiliae*) and the galls are widespread in the UK. The hollow structures are produced in response to the mites feeding on the underside of the leaf. The mites subsequently enter the galls to shelter, feed and breed. This particular mite is restricted to lime (or linden) trees. Similar galls are often produced on alder leaves by a related mite – *Eriophyes laevis*.

Myles O'Reilly
Scottish Environment Protection Agency
Glasgow, UK

⁇ Plant depression

When I was growing up in Luanda, Angola, I remember a peculiar bush. When you touched it, the area where contact was made wilted immediately, with all the leaves on that branch drooping. This lasted for only 2 or 3 minutes, after which the leaves slowly became erect and returned to normal. Does anyone know the name of this bush and why it behaves in this way?

Luis M. Luis
Virginia, US

My botanist father offers a theory as to why *Mimosa pudica*, the drooping plant your correspondent observed, behaves like this: the ploy discourages herbivores. If a herbivore starts chomping at the leaves and the plant immediately starts to wilt, pretty soon it will look rather unappetising. Presumably a lush bush appears to contain more nutrients than one that is wilting, so the herbivore moves on to lusher-looking plants instead. This seems a rather neat defence mechanism.

Chris Wright
Hampton East, Victoria, Australia

The peculiar bush which wilted when touched was almost certainly the 'sensitive plant', *Mimosa pudica*. This is a spindly tropical shrub with delicate feathery leaves, which fold up very quickly if the plant is touched or shaken. Even in colder climates you can grow it easily from seed. It makes an unusual houseplant with attractive pink fuzz-ball flowers, but the main attraction is, of course, its sensitive nature. The plant will survive outside in warm weather, where a sudden gust of wind or a few raindrops will cause the whole plant to close down.

Presumably this sensitivity evolved as a defence against damage by tropical rainstorms. On the other hand, in breezy conditions the plant loses its sensitivity because a breeze, unlike the touch of an animal, holds no threat. The plant then refuses to react even to a good shaking.

The plants are certainly efficient photosynthesisers because they take only a few months to grow to a height of around 1 metre even in the UK, but the leaves are not robust and would easily be damaged if rigidly attached.

A leaf consists of four or six fronds, each about 6 centimetres long and attached to a thin stalk. Each frond carries an average of 17 pairs of small leaflets, which are the first to react when the leaf is touched. If left undisturbed for a few hours in hot and humid conditions, the plant becomes almost hypersensitive as the warmer conditions are more conducive to sensitivity, so touching just one leaflet will cause the whole leaf to progressively close up (see photo).

With choreographic accuracy the leaflet pairs close in sequence along one frond, then the leaflets on the remaining fronds follow suit. With each leaflet pair taking about half a second to react, the fronds are fully closed in around 15 seconds. Then the fronds close like the fingers of a hand and, after a pause, the leaf stem itself angles downwards when a thickened area close to the stem gets the message. Sometimes other leaves on the same stem will react in a half-hearted sort of way.

At a rough estimate, the message from the initial touch point travels at about half a metre per minute. This is not much by mammal standards but good going for a snail and definitely a lightning reaction for a plant. If this isn't evidence for the presence of nervous systems in plants then I don't know what is. Does anyone know how it carries it out?

John Rowland
Derby, UK

Many of the species in the *Mimosa* genus are sensitive to touch. It is native to Brazil but can be found in most tropical regions of the world now. *M. pudica* is remarkable in its genus for being seismonastic, meaning it is sensitive to vibrations and also to touch. Once touched the leaflets move rapidly together, closely followed by the whole leaf collapsing.

This is caused by a reaction in the cells of the thickened tissue, or pulvini, at the base of the leaflets and leaves. These cells are sensitive to internal or turgor pressure. When a leaf is touched an electrical signal causes this turgor pressure to drop rapidly and the pulvini collapse – the leaves follow the leaflets because they are connected. The actual mechanism for this movement is still not fully understood. There is evidence that a simple turgor mechanism may be an oversimplification. More study is needed.

Peter Scott
Plant conservationist
School of Life Sciences
University of Sussex
Brighton, UK

❓ Lunatic cactus

My Cereus forbesii *cactus flowered last night (see below), coinciding with a full moon. The* Selenicereus grandiflorus *cacti that I had in Bangladesh always flowered at or within a couple of days of a full moon or, more occasionally, a new moon. How is flowering in such plants triggered by the lunar cycle?*

Hugh Brammer
By email, no postal address supplied

Plants like *Selenicereus* flower at night, when temperatures are low and the creatures that pollinate them are about. A white flower opening at night is highly visible, particularly with a full moon to illuminate it, so nocturnal flowering makes sense in evolutionary terms.

There is also good evidence to suggest that plants sense the length of the night-time and that these periods trigger flowering. Therefore 'interrupting' the night with a bright light such as a full moon could have an effect on flowering in plants. But

I know of no scientific studies that have shown this to be the case.

An internet search provides very few reports of *Cereus* or *Selenicereus* flowering in response to a full moon. So the questioner's observation is likely to be the result of chance.

There are 28 days in a lunar month, and on three of those days the moon will be at its brightest (approximately a full moon). So the plant has roughly a 1-in-9 chance each month of flowering at the time of the full moon. I suspect that if it flowers at another time, the observation is not considered remarkable and so does not get reported.

Peter Scott
Plant conservationist
School of Life Sciences
University of Sussex
Brighton, East Sussex, UK

I have a specimen of *Selenicereus grandiflorus* in my greenhouse which produced seven flowers during the summer of 1998. The first flower opened in the night of 20 or 21 June and the others at regular intervals during the following two weeks. The last one opened on the night of 5 July. There was a new moon on 24 June and the first quarter was on 1 July. So it is difficult to conclude that the flowering of my cactus was triggered by either the full moon or a new moon.

However, the question aroused my interest and I did some research. I discovered the work of Yosef Mizrahi at the Ben-Gurion University of the Negev in Israel. He and his team have been exploring the possibilities of growing the vine cacti *Hylocereus* and *Selenicereus* as fruit crops: his team have more than 240 genotypes of vine cacti from these two genera in their gene bank.

I emailed the above question to him. His response was that while these species flower at different times of the year, he and his team have not observed that the full moon triggers the opening of their flowers, although he admits that his researchers have not been actively looking for such a phenomenon.

Trevor Lea
Oxford, UK

'…there is no greater folly than to be very inquisitive and laborious to find out the causes of such a phenomenon as never had an existence, and therefore men ought to be cautious and to be fully assured of the truth of the effect before they venture to explicate the cause.'
    The Displaying of Supposed Witchcraft by John Webster (1677)

I have been growing cacti for more than 60 years and have heard this claim several times. I would refer those who asked to this quote from John Webster.

Although many flower at night, I have observed no examples of any cactus coordinating its flowering with phases of the moon. I have digital photographs of hundreds of images of cacti in flower, some of which have been published in journals. They were taken with a digital camera and are automatically labelled with the date. Having just checked these flowering dates against the phases of the moon, I can tell you that there is no correlation, hence the quote I provide above.

The water loss from large flowers is extreme and so cacti generally keep theirs open for very limited periods. *Micranthocereus purpureus*, for example (see my photograph on the previous page, taken at 7.15 am on 2 February 2007), opens its flowers after sunset and closes them permanently the next morning, before sunrise. The flower shown is near to closing. Incidentally, that one missed the full moon by four nights.

Jim Ring
Nelson, New Zealand

❓ Stem well research

Some deciduous trees of the Combretaceae family, such as Terminalia tomentosa, *show a remarkable ability to store water in their stems during the dry season in India. A small cut in the tree shows the amount they can carry. How do they do this and do any other plants or trees store water in a similar way?*

R. Shyama Prasad Rao and H. L. Prabhakar
Mysore, India

The picture shows a stem section of *Terminalia tomentosa*, which when cut produces a steady stream of fluid. The plant does this by acquiring water during the rainy seasons and holding onto it in cells known as parenchyma cells. These cells are undifferentiated and can expand to accommodate fluid throughout the stem and roots. The parenchyma cells are immediately adjacent to the xylem of the tree, so making contact with this fluid-carrying material is easy. The tree does this to cope with water shortages during the dry season. The tree will always take up water when it is available.

Other trees do this too, as do some vines. They all use the same mechanism. In fact some vines in South America and Africa store enough water to supply native people when surface water is not readily at hand.

Trees in temperate regions also collect water. Members of the white oak family store water from autumn to spring and use it during the dry summer months. Other trees that do this are the famed sugar maples of north-eastern America, black birch, negundo maples, Norway maples, and many species of grape vine. If any of these trees are tapped in early spring, the water flow with affiliated nutrients and sugars can be quite prominent, and is the starting point for maple and birch syrup. If grape vines are cut during spring or early summer they will often bleed so copiously that the plant can dehydrate and die. That's why there are special times of year for vine pruning.

The temperate trees inherited this trait from tropical species, although they use it differently. For them the aim is to withstand the rigours of cold weather rather than to avoid dry-season dormancy. It is the tree version of being a cactus.

H. William Barnes
Warrington, Pennsylvania, US

❓ Tree love blossoms

While walking in the New Forest in southern England, my wife and I came across two trees that appeared to be holding hands (see below and overleaf). They even had a branch linking their trunks. How and why did this happen? A hopeless romantic I may be, but even I don't think it's because the trees are in love.

John Badman
London, UK

The phenomenon is called inosculation. It is not uncommon between trees of the same species but much rarer between different species. It seems to be most common when a branch or trunk interpenetrates another within the same tree, but it can occur between neighbouring trees. Two branches rub together, removing some of the bark, and then a natural graft forms and the two trees grow together. I have seen a single branch joined to two or three others in an overgrown coppice.

Trees and shrubs that will readily inosculate include elm,

holm oak, live oak, golden oak, olive, pear, apple, peach, almond, beech, hornbeam, linden, hazel, crepe maple, dogwood, golden willow, wisteria, grape, privet, laburnum and sycamore.

When inosculation is created by human intervention it is called pleaching, and is commonly used to form an avenue of trees all joined together (see www.orchardsedge.com/articles/ pruning_advice/pleaching).

Other examples include the 'tree circus' (opposite, at top) created by American Axel Erlandson which opened in Scotts Valley, California, in 1947 (www.arborsmith.com/treecircus. html), while American artist Dan Ladd creates what he calls living sculptures using pleaching (www.danladd.com/exam-ples.html). Pleaching can even be used to construct living build-ings (opposite, at bottom): www.archinode.com/biena102.html

Nick Kotarski
East Bergholt, Suffolk, UK

There is a similar but more advanced twinned tree near Strachan in Aberdeenshire at approximate UK Ordnance Survey grid reference No 637927 (see overleaf). A friend who has observed the trees growing for many years tells me that the recipient tree used to be much the smaller of the two, but its growth has now outstripped that of the donor tree.

Maureen Young
Aberdeen, UK

Pleaching is an ancient agricultural and horticultural technique that involves weaving together branches to form walls, ceilings, arches and other ornamental structures that can, in extreme examples, form homes. Invented by horticulturists during Roman times, the technique is mentioned in Shakespeare's Much Ado About Nothing *and is still practised in gardens today, where it is once more gaining interest among creative landscapers. The method is well suited to walls, fences and walkways and often involves training branches to grow alongside other horizontal branches off the side of the main trunk.*

In its purest form, pleaching involves making small cuts to encourage branches to graft to each other in a kind of forced inosculation. Trees that self-graft are, naturally, the most suitable species. These typically have pliable branches that can be directed into and through cuts in other branches to form load-bearing structures. Apple, linden, elm, hawthorn and pear trees work especially well. Branches are usually bent rather than cut. And where the limbs touch, they join, giving strength to the tree.

Mount Vernon in Virginia, former home of the first president of the United States, George Washington, has fine examples of pleaching – Ed.

❓ Pigment pattern

I found this leaf in a park in Oslo, Norway. What is the explanation for this pattern of green pigmentation?

Fredrik Størmer
Oslo, Norway

The leaf shown is typical of a hazel leaf in autumn. At this time of year, chlorophyll is broken down so that its components can be recycled by the plant. The extremities of the leaf are exposed to the greatest variations in temperature, so chlorophyll breaks down there first. This reveals the underlying yellow carotenoid leaf pigments, hence the edge of the leaf is yellow with splashes of orange-brown as the leaf cells finally die. This region will spread inwards until the whole leaf is brown or the leaf falls off the tree.

Over the major veins there is only a very thin layer of photosynthetic cells, which means their chlorophyll is broken down early compared with other parts of the leaf. In contrast, the green stripes that appear between the veins are areas where the leaf is thickest and the photosynthetic cells are protected for longer from the changes in environment. In this case, the layers of cells near the leaf surface have probably already turned yellow but this is masked by the cells below that still contain chlorophyll. The green patches look especially dark but this is probably an optical illusion resulting from the surrounding yellow tissue.

No two leaves are in identical condition as they die in autumn, because of the differing microclimates experienced by each leaf, and thus the development of pigment patterns is remarkably variable. Never expect to see the same pattern twice.

Peter Scott
Hove, East Sussex, UK

6 Bugs and blobs from the deep

? Thingies

Swimming across Coogee Bay in New South Wales one morning, we came across thousands of strange creatures similar to the one shown here, floating at depths down to about 2 metres. They were hard but also flexible, with water inside and a small hole at one end. Their length varied from about 3 to 30 centimetres and their walls were between 2 and 5 millimetres thick. Their skin was marked with many small protrusions, the size of which varied from one creature to another. Unlike jellyfish they appeared to be completely harmless. No one I've spoken to from the area has ever come across anything like this. So what were they and why were they there?

Philippe Wilmotte
Maroubra, New South Wales, Australia

The tubular object is a pyrosoma. These are colonial tunicates, related to sea squirts, salps and doliolids. Each tube, or tunic, is a leathery gelatinous matrix that contains a number of individual tunicates, or zooids. One end of the tube is blocked and the other open but guarded by a controllable diaphragm. The inner surface of the tube is very smooth. In contrast, the tube's outer surface is rough because each zooid projects out of it to feed.

Each pyrosoma is made up of numbers of individuals buried in a common, tubular, gelatinous matrix. Each zooid is a filter feeder, pumping seawater by ciliary action from outside the tube and passing it through a pharyngeal branchial 'basket', or gill, where planktonic food items and oxygen are extracted. The seawater is then expelled to the interior of the tube, before leaving via the tube's open end. When all of the zooids in the tube are pumping water into the interior the pyrosoma moves by jet propulsion. Pyrosomas vary a great deal in length, from a few millimetres to 30 metres. They undertake vertical migrations, tending to be at the surface at night and at depth during the day.

The name pyrosoma translates as 'fiery body' and the colonies show intense, sustained bioluminescence when stimulated mechanically or by light. Each zooid has two light organs that contain luminescent bacteria. A lit-up colony can be seen from up to 100 metres away in clear waters in the middle of the night.

Bioluminescent light from one colony will stimulate another to flash. Light output is always preceded by the cessation of ciliary pumping, so the lightshow is generally believed to warn of poor food supply or the presence of predators.

Pyrosomas are entirely harmless and occur in swarms in productive areas of the world's oceans. As with jellyfish, swarms sometimes drift into coastal shallow water. They have another common feature with jellyfish: pyrosomas are

94 per cent water so represent a low-value diet. However, their best-known predator, the leatherback turtle (*Dermochelys coriacea*), still eats enormous quantities, probably hunting them at night.

John Davenport
Professor of Zoology
University College Cork, Ireland

I had put this question to the back of my mind until I started researching gelatinous zooplankton for a lecture I am preparing. This generated a eureka moment.

I think the creature in the photo is the thaliacean *Pyrosoma atlanticum*. At first glance it looks like a sea cucumber, or holothurian, but the fact that many specimens were found in the water column and your hint that the body is gelatinous (hard but flexible) gave me doubts.

Despite having a similar gelatinous constitution, thaliaceans are not directly related to jellyfish, which explains why they do not contain cnidocytes – the stinging cells that provide the nasty sting experienced from some jellyfish species.

P. atlanticum is a colonial species made up of zooids gelled together in a gelatinous tunic, which gives them a 'bumpy' appearance. They can be pink, as shown, and have one hole at the end of the tube. They are often found in swarms, as the questioner describes. These swarms sink rapidly when dead and have been found to accumulate in patches on the deep-sea floor, creating an important source of fresh organic carbon for deep-sea animals.

Tania FitzGeorge-Balfour
Queen Mary University
London, UK

❓ Get your skates on

I recently saw this collection of pond skaters on our garden pond. Can anyone tell me what they were doing and why they adopted this strange formation?

Dominic Cox
Enfield, Middlesex, UK

Such formations are fairly common. Gerrids (pond skaters) have the sucking mouthparts typical of true bugs. They generally feed on insects drifting on the surface of the water; usually their prey are injured or non-swimmers. The pond skaters in the photograph are feeding on a partly submerged insect, perhaps a fly.

Gerrids pierce their victim, inject digestive juices and suck it dry. They can read surface waves with exquisite sensitivity: a fly buzzing helplessly in the water acts as a magnet for every nearby opportunist. In nature, the lightly built gerrids have to move fast because other creatures – whirligig beetles, for example – prey on similar victims and are faster and more robust.

When food is in short supply, any tempting prize attracts every gerrid within call. Often there is a scrum, and the prey soon has more perforations than a pincushion. When more than one attacks the same food item, the diners generally form a ring because gerrids are not averse to cannibalism, so it does not do to let one's fellows climb on top.

Jon Richfield
Somerset West, South Africa

Pond skaters, or water striders, are predators and scavengers. They eat mainly terrestrial insects and spiders dropped and trapped at the water surface. Pond skaters cling to their prey with their short front legs and inject digestive fluids into it via their proboscis before sucking out the resulting soup of body fluids.

Typically a feeding bout lasts about half an hour. If the prey is large enough, it can attract several pond skaters to eat at the same time. I suspect that in the centre of the star-like formation of pond skaters is a big fat fly.

Matti Nummelin
Department of Biological and Environmental Sciences
University of Helsinki, Finland

Thanks to Tim Gossling and Dave Challender for sending in photos of similar events. You can see Dave's photo, complete with a wasp in the middle, at bit.ly/1sOGpw – Ed.

❓ On the March

In March 2010, I came across the line of caterpillars shown in the photo below. They were on the seafront road in Carnac in north-west France, heading towards the sea, maintaining close head-to-tail contact. The only vegetation between them and the sea was some raised flower beds. Can anybody identify these caterpillars and explain their behaviour?

Bill Richardson
Johnstone, Renfrewshire, UK

These are unmistakably the caterpillars of the pine procession-ary moth, *Thaumetopoea pityocampa*. In France this species was only found south of the Loire valley until quite recently, but it is extending its range northwards as the photo confirms. It has not been seen in the UK so far, although its close relative the oak processionary moth, *Thaumetopoea processionea*, has recently established itself there.

I don't think anyone knows exactly why these caterpillars

move in columns, but it is probably because there is safety in numbers, especially when the caterpillars possess quite a nasty weapon. Each segment of the caterpillar's body has a dorsal pouch bearing microscopic hair-like barbs containing a toxic protein, thaumetopoein. If you handle the caterpillars and get some of the barbs stuck in your skin, you are very likely to scratch, the action of which releases the toxin. Normally this is no more than an irritant, but in sufficient quantity it can produce dizziness and even anaphylactic shock. It can also cause widespread cell death in the affected area, making it effectively fatal for a dog to try licking the caterpillars. Rather than have to remove the dog's tongue, a vet will usually put the dog down.

In my part of south-west France the moth starts laying eggs in July and the caterpillars can reach maturity by Christmas. It seems that in Carnac it takes them a few months longer. When fully grown the caterpillars form columns of the kind depicted and march off to find a suitable spot to burrow into the ground and pupate. There doesn't seem to be a natural leader – whichever caterpillar finds itself at the head of the column determines where they will go. If you join the head of the column to the tail, the caterpillars will quite happily march around in circles.

It seems the caterpillars are preprogrammed to march for a certain time. The ones I have bred march around their containers over a period of several days, stopping periodically before finally pupating. In the wild this is clearly a form of migratory behaviour, ensuring that the population does not remain confined to the tree where they hatched.

Each brood of caterpillars lives communally in a silk nest that is usually attached to the new growth on pine trees. They feed by night, not just on pines but also on other conifers such as firs and cedars, retreating to the safety of their nest during the day.

In France the natural range of pine trees is restricted to mountainous areas, with the exception of the Aleppo pines on the country's Mediterranean coast. Wild pines grow slowly and their needles produce defensive tannins, making it hard for processionary moths to develop. In mountainous areas it may therefore take several years for a processionary moth caterpillar to grow to maturity.

Where pine trees are cultivated commercially, such as in the vast pine forest that is the Landes de Gascogne park, the trees grow rapidly and the fresh growth is easily digested by the caterpillars. It is not unusual to see pines completely defoliated and even killed by caterpillar infestations. Living close to such an outbreak is uncomfortable. The air is filled with their barbs, which glint in the sunshine, and life becomes one perpetual itch.

The oak processionary moth can likewise defoliate oak woodland. This does not happen in southern France, where I believe the natural diversity of species keeps them in check, but there have been infestations on an epidemic scale in northern France and Belgium. Foresters in the UK fear that similar outbreaks could devastate the country's oak forests and that trees already menaced by diseases such as sudden oak death might be especially vulnerable.

Processionary moths are not immune from predators. Some birds, such as crested tits, cuckoos and hoopoes, will eat the caterpillars; the eggs, larvae and pupae are attacked by parasitic insects; and the adults are taken by bats.

Terence Hollingworth
Blagnac, France

? Fly trap?

When my family returned from a two-week summer holiday, we found a cluster of dead flies stuck to our porch window similar to the ones in this photograph. Each was surrounded by a hazy mist on the glass. We have lived in our house for 20 years and have never seen this before. Even after two weeks, nothing had changed. The flies hadn't moved and no predator had come to eat them. Can anyone explain what has happened?

Stephen Ryder
UK

The flies are almost certainly affected by the parasitic fungus *Entomophthora muscae*, which infects and kills domestic flies, *Musca domestica*. Infected flies are forced to land while the fungus grows inside the abdomen, causing it to swell. Finally the fungus breaks out and causes the mist-like structures seen on the window.

Infected female flies with swollen abdomens are especially

attractive to male flies, which try to mate with what they think are fecund females. The disease then spreads from female to male, and the greater the population, the faster it spreads.

David Fleet
Süderstapel, Germany

The flies show typical symptoms of infection by an entomo-pathogenic fungus. The white halo around each dead fly is a deposit of the forcibly discharged infectious fungal spores.

When one of these spores touches a live fly it germinates and penetrates the insect's cuticle. The fungus then grows internally until the fly's body is full of fungal cells. The infected fly becomes sluggish, comes to rest and dies. If the humidity is high enough the fungus then grows out of the intersegmental membranes, incidentally gluing the dead insect more firmly to the windowpane, and fires off a fresh round of spores.

Some fungal species alter the behaviour of the host insect, causing it to die in an unusually elevated position, which improves the chances of a spore hitting a fresh host. A striking example is *Entomophaga grylli*, which infects grasshoppers and causes 'summit disease', where the infected insects climb to the tops of grass stems to shower spores on the population below.

Under humid conditions, entomopathogenic fungi can cause impressive epidemics in insect populations. The very wet British summer of 2008 would have been favourable for infection, and I found several infected flies in my garden.

Chris Prior
Bampton, Devon, UK

❓ Something eats wasps

I photographed this amazing sight in Croatia in July 2007. In light of New Scientist's *book* Does Anything Eat Wasps? *I'd like to know what's going on here. Which insect is eating which, and is it common?*

Richard Garner
UK

The smaller insect is a wasp (probably a potter wasp) and the other is a robber fly (Asilidae, probably of the genus *Mallophora*).

Robber flies can grow to more than 2 centimetres long and are quite fearsome in appearance, with the orange tufts of hair around the face thought to protect it from its prey. This one is a wasp mimic and, as we can see, a killer.

They are known to attack large insects in flight, gripping the prey with their forelegs and then piercing it with their proboscis to inject a neurotoxin along with enzymes that break down proteins. The robber fly then lands and sucks the liquidised juices straight from the body of its prey. In this picture the wasp is in the process of being paralysed.

There are more than 7000 different species of robber fly. Some are fairly common, and few are as impressive as the one shown.

Peter Scott
School of Life Sciences
University of Sussex
Brighton, UK

Both insects are from the order Diptera. The larger is a robber fly of the family Asilidae, which has caught a thick-headed fly of the family Conopidae. The prey will be sucked dry by the attacker's piercing mouthparts. The victim is not a wasp, but a wasp-mimicking fly. You can tell, as they both have only two wings (hence di-ptera) while wasps have four.

P. H. van Doesburg
Emeritus entomologist
Natural History Museum
Leiden, Netherlands

The picture shows a large robber fly, or asilid, eating a wasp. Apart from being a beautiful specimen of an impressive species, this wasp-eater is doubly interesting in that it seems to mimic a large spider-eating wasp from the Pompilidae family.

Asilids are better known for mimicking bumblebees. So why is this one mimicking a wasp? Pompilids are well behaved, but the large ones sting agonisingly if grabbed. So to mimic a pompilid offers a broad hint of danger. In fact, a stab from the beak of any large asilid is memorable, so if you are not an entomologist just admire them from a distance.

Most asilids sit on rocks or vegetation, ambushing passing food such as flies, butterflies, other asilids or, indeed, wasps. Some even rip spiders from their webs.

This behaviour is not as common as it once was. As wild

territory shrinks, the loss of such species and their interrelation-
ships is beyond the imagination of ecological incompetents. As
a child I took asilids for granted. Nowadays, to see a nice big
asilid is a rare treat.

Jon Richfield
Somerset West, South Africa

A few years ago my family, my friend and I visited a place called
the Otter Pool on the Raider's Road in Dumfries and Galloway
in Scotland. As we sat eating our picnic in the summer sunshine,
a dragonfly landed on my friend's shirt sleeve. It had caught a
wasp and was happily munching away.

The dragonfly stayed on my friend's sleeve for around 5
minutes, until the entire wasp was consumed. My father filmed
the event and you can see it at www.tinyurl.com/6qmne2.

Mark Jepson
By email; no postal address supplied

The lowly wasp certainly has its place in the food chain. Indeed,
the question should possibly be 'what doesn't feed, in one way
or another, on this lowly and potentially dangerous insect?'

Here are a few that do, the first list being invertebrates:
several species of dragonflies (Odonata); robber and hoverflies
(Diptera); wasps (Hymenoptera), usually the larger species
feeding on smaller species, such as social paper wasps (*Vespula
maculata*) eating *V. utahensis*; beetles (Coleoptera); and moth caterpillars (Lepidoptera).

The following are vertebrates that feed on wasps: numerous species of birds, skunks, bears, badgers, bats, weasels, wol-
verines, rats, mice and last, but certainly not least, humans and
probably some of our closest ancestors.

I have eaten the larvae of several wasp species fried in butter,
and found them quite tasty.

Orvis Tilby
Salem, Oregon, US

The definitive source on European birds, *Birds of the Western Palearctic* published by OUP, lists a remarkable 133 species that at least occasionally consume wasps. The list includes some very unexpected species, such as willow warblers, pied flycatchers and Alpine swifts, but two groups of birds are well-known for being avid vespivores. Bee-eaters (Meropidae) routinely devour wasps, destinging them by wiping the insect vigorously against a twig or wire. And honey buzzards raid hives for food. They are especially partial to bee larvae, but in the UK, wasps – again mostly larvae – also form a major part of their diet.

Simon Woolley
Winchester, Hampshire, UK

This photograph, taken in my garden, shows a mason wasp having its internal juices removed via the proboscis of a large insect.

Tim Hart
La Gomera, Canary Islands, Spain

In July 1972, I was snorkelling off the Californian island of Catalina when I saw a crab in a crevice holding a wasp, which was still moving. The crab held the wasp in its right pincer and used the left to manipulate the wasp's abdomen to its mouth. The crab did not show any sign that it was startled by the taste of its meal.

Garry Tee
Auckland, New Zealand

Badgers will dig out a wasps' nest and eat the larvae and their food base. The picture below shows an underground nest demolished in the summer of 2003.

Tony Jeans
Cheltenham, Gloucestershire, UK

I was once idly observing a wasp crawling round the edge of a waterlily leaf in my pond when it paused to drink. There was a sudden flurry of activity as a frog leapt from its hiding place and swallowed the wasp.

The frog did not appear to suffer any ill effects, so I captured another wasp, tossed the hapless creature into the pond and waited. The frog was slow on the uptake, but there was another disturbance in the water and this time a goldfish snapped up the wasp. The fish, too, seemed undisturbed.

My curiosity now thoroughly aroused, I wondered whether the fish could be induced to consume further wasps. For the next hour or so I continued to hunt down luckless wasps and throw them into the pond. Some got away, some were eaten by the fish, and a few were swallowed by the frogs.

John Crofts
Nottingham, UK

Returning home late one night I heard the persistent buzzing of a wasp in the kitchen window. It appeared to be struggling around at the bottom of the window, unable to fly properly. A tiny red spider was attached to the underside of its abdomen. The wasp must have been some 20 times larger than the spider, but the spider was positioned where the wasp was unable to mount a counter-attack.

The next morning revealed an empty, transparent wasp exoskeleton.

John Walter Haworth
Exeter, Devon, UK

You may be interested in these photos (see overleaf) I took in my garden in Melbourne in April 2007. They show conclusively that praying mantises eat at least the heads of European wasps: one photo shows the predator sucking up the wasp's brain before cleaning itself.

I did not see the mantis eat any of the wasp's body. It left it behind, but I didn't see whether it did so on purpose or dropped it accidentally.

Wendy Kimpton
Malvern, Victoria, Australia

Accompanying the excellent pictures of the praying mantis eating a European wasp in Melbourne, the contributor suggested that the mantid has sucked up the wasp's brain and discarded the remainder. In fact, the mantid is eating the last piece of flight muscle from the thorax – the tastiest and most muscle-bound part of the wasp – ideal tucker for a hungry mantis.

The head was probably discarded soon after the wasp was captured, as the abdomen would be once the nutritious contents of the thorax were devoured.

Philip Spradbery
Canberra, Australia

The writer is the author of Wasps, *published by Sidgwick & Jackson* (1973) *– Ed.*

❓ Dune bug

I found this insect, and others like it, in my friend's home near Stamford, Lincolnshire, in eastern England. It looks as if it is covered in sand but this is clearly a form of camouflage. Interestingly, there were no obvious sand deposits nearby. What is it? And why does it look like this?

Ian Richardson
London, UK

The photograph is of a larva of the true bug *Reduvius personatus*. Although many insects are referred to as bugs, only the insects in the order Hemiptera are true bugs. This particular species is noted for its habit, while in its larval form, of coating itself with dust and any other debris that it can find as a means of camouflage. This activity begins as soon as the larva emerges from its egg and is repeated every time the larval skin is shed as the insect grows – five time in total. The adult does not cover itself with dust, but is dark brown and fully winged.

The insect is associated with mammal habitats, especially those of humans, where it is active mainly by night and so generally escapes notice. It is a predator that feeds on other invertebrates, such as silverfish, by grabbing them with its legs and inserting its mouthparts into their bodies to relieve them of their nutritious innards. The mouthparts of true bugs consist of a tube-like instrument, the rostrum, which is one of the features that distinguish them from beetles, with which they are sometimes confused.

R. *personatus* is found across Europe and also in North America, where its camouflage technique has earned it the common name 'masked bug' or 'masked bedbug hunter'. In the UK it is sometimes known as the 'flybug'. The flybug is not as common as it once was and perhaps this reflects the decline in abundance of a favourite prey species – the bedbug *Cimex lectularius*. Recently, however, populations of *C. lectularius* seem to be increasing so it is possible that the flybug may yet make a comeback in the UK.

Ray Barnett
Bristol, UK

Thanks to Søren Tolsgaard, a curator at the Aarhus Natural History Museum in Denmark, for this photo of a camouflage-free adult R. personatus – Ed.

❓ Strings attached

While on holiday in Greece, we found some strange eggs in the sea (see below). They were composed of a jelly-like substance with an embryo clearly visible. Whatever laid them was obviously quite big because the eggs formed long strings similar to toad spawn. The eggs pulsated if they were taken out of water. Can anyone identify them?

David Castle
Manchester, UK

These are not actually eggs at all. The picture shows a gelatinous creature called a salp. These pelagic tunicates are often barrel-shaped and at least partially hollow. Believe it or not, they belong to the chordate phylum, the same as humans.

The questioner mentions that the 'eggs' pulsated when lifted out of the water. This means they were holding a live salp. If you spotted one in the water – say when snorkelling – you would be able to see that the salp moves by contracting its main body walls and pumping out water. The opaque part in the photograph described as the 'embryo' is in fact the salp's digestive and primitive nervous system. Salps can be free individuals or form strings (see photo below) – some species even create huge mats or large hollow bodies many metres long which almost replicate their individual body shape on a giant scale.

Salps feed on plankton and are common in most waters, but more so in temperate and equatorial climes. They reproduce very quickly by budding off clones as a direct response to plankton blooms. Often, after plankton blooms have faded, dead or dying salps wash up on beaches or form tidal bands before sinking to the sea floor.

Dave Banks
Wellington, New Zealand

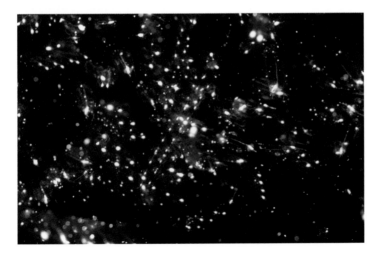

Salps are pretty cool. Picture a transparent gelatinous barrel, open at both ends, with ring-like muscle bands and a primitive proto-spinal cord (known as a notochord). They are a pelagic version of sea squirts, or tunicates, meaning that they spend their whole life high up in the water rather than stuck to the bottom. They move swiftly by jet propulsion.

The 'long strings similar to toad spawn' that are described are the colonial part of the life cycle, where a nurse mother clones off many juveniles that remain stuck together while they grow.

They may look like jellyfish, but they are members of our phylum – a sort of a strange not-quite-vertebrate-not-quite-invertebrate link in the tree of life.

They are vegetarians, filtering plankton from the water, and are often highly bioluminescent (providing unlimited entertainment for bio-nerds like me). They are also the primary biological indicator that we use in Australia to predict and detect the presence of a type of dangerous jellyfish called an irukandji. So, these gelatinous vegetarian barrel cousins of humans also help us to keep our beaches safe.

Lisa-Ann Gershwin
State Marine Stinger Advisor
Brisbane, Queensland, Australia

？ Killer bush

While out on a country bike ride, I saw this unfortunate insect impaled on the thorn of a low bush. We'd had strong winds in the days beforehand and I can only assume that the insect was blown onto the thorn, which has penetrated the open wing casing before impaling the body. What are the chances of an event like this occurring?

Paul Worden
Portland, Victoria, Australia

The beetle is *Phyllotocus macleayi*, a nectar scarab common in south-eastern Australia, often found in large numbers feeding on the blossoms of eucalyptus trees. The plant is sweet bursaria (*Bursaria spinosa*), sometimes called prickly box or blackthorn.

Forensic interpretation of this photograph is complicated.

Curiously, the right mid-leg of the beetle is detached from the body and hanging precariously from the left hind leg. The long tarsi, or feet, appear to have been partially broken off and the hole in the elytron, or wing cover, made by the thorn appears to be bigger than the maximum diameter of the thorn at its base. This suggests the beetle sustained further damage after it was impaled.

The thorn penetrates the beetle in the same place as an entomologist would insert a pin, and sometimes a badly placed pin will break off the aforementioned right mid-leg. It would be very difficult to break the elytron in a similar way by manipulating a pin once inserted.

Examples of impalement of other species of scarab beetles on spines or other sharp parts of plants and on barbed wire have been recorded in the literature in Australia, but only rarely.

It is very unlikely that the wind alone could have impaled the elytron. That would be like an insect collector attempting to pin a beetle by throwing it at the pin. The elytron is part of the hard exoskeleton and would almost always deflect the glancing blow of a pin or spine. The chances of impalement through an elytron while in flight would appear to be very remote because the flying beetle holds its elytra open, at a wide angle to the body, and they would hinge back towards the body if touched on the outer side by a spine.

A more likely scenario is that the strong winds blew down a twig or branch to which the beetle was clinging and that the extra momentum of beetle plus branch impaled it. Dislodgement of the fallen plant material at a later date could account for the other damage. Similarly, strong winds might have caused one sweet bursaria branch to thrash against another on which the beetle was clinging.

Ian Faithfull
Extension Support Officer
Catchment and Agriculture Services
Carrum Downs, Victoria, Australia

Dung beetles frequently impale themselves on the barbs of wire fences while in full flight. The picture below shows the typical place that barbs catch the insects. About 2 metres away from where I took this picture, another dung beetle had been impaled in the same place.

Toshi Knell
Nowra, New South Wales, Australia

Actually, this event is quite likely in certain areas. The poor beetle probably did not get there by chance, but rather because it was put there.

In Victoria, the culprit is likely to have been a grey butcher-bird (*Cracticus torquatus*). These predatory birds, which are about the size of a small dove, eat large insects, small mammals and other birds. They skewer their prey on thorns to hold it while they eat. Sometimes they will impale the prey and leave it for a snack later. Suitable bushes near nesting sites can be festooned with victims, including poultry chicks.

There is a record of a butcher-bird returning to its nest to find its three chicks dead after a spell of cold rain. The bird took them from the nest and hung them in its nearby larder, returning to eat them a few days later.

Shrikes, which are common throughout Eurasia, Africa and North America (and also sometimes known as 'butcher-birds', although they are not related), have similar habits.

Rob Robinson
British Trust for Ornithology
Thetford, Norfolk, UK

❓ Finding Nemo's flies

I was walking outside during a recent heatwave in Sydney, swatting all the newly hatched flies and contemplating how much of the Earth's biomass is accounted for by insects, when it occurred to me that I have never heard of any insects that live in or on the sea. Are there any?

Rob Moore
Sydney, Australia

The only truly marine insects are the sea skaters of the genus *Halobates*, order Hemiptera. Lanna Cheng gives a full account of these in the *Oceanography and Marine Biology Annual Review* of 1973 (vol 11, p 223), while www.ices.dk/products/fiche/Plankton/SHEET147.PDF provides online information. This is updated in a 2004 paper: 'The marine insect *Halobates* (Heteroptera: Gerridae): biology, adaptations, distributions and phylogeny' (vol 42, p 119).

Alistair Lindley
Sir Alister Hardy Foundation for Ocean Science
Plymouth, Devon, UK

Marine insects certainly exist, although they constitute only a tiny percentage of the total number of insect species. The species range from those which live in brackish water, for example in saline lagoons and rock pools at the upper limit of the splash zone, to species which live on or beneath the sea surface.

A survey of brackish lagoons in Ireland recorded 77 species of insect, although most of these are not confined to the habitat. Flies (Diptera), beetles (Coleoptera) and true bugs (Heteroptera) were the commonest groups. The flies are usually present as larvae, the others in both the larval and adult stages.

True marine insects can be divided into species that live

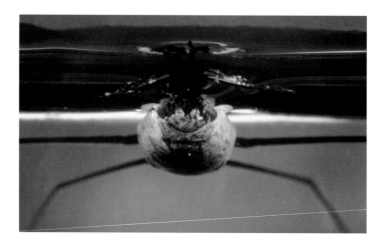

their entire lives on the sea surface and species that can tolerate immersion either as adults or as larvae.

The five species of sea skaters (*Halobates*) are examples of the first group and are often cited as perhaps the most extreme example of marine insect (see photo above). It is true they never voluntarily venture onto land, but they actually live only on the water surface without getting wet – just like pond skaters and

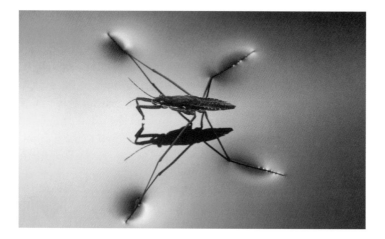

water striders (see photo opposite, at bottom), their freshwater fellow members of the family Gerridae. *Halobates* lay their eggs on flotsam and jetsam. They can be found in all the world's oceans between the latitudes of around 40 degrees north and south.

Arguably, species such as the true bug *Aepophilus bonnairei*, an inhabitant of rocky Atlantic shores of western Europe and north Africa, are more truly marine because they can survive regular and prolonged immersion in seawater.

At high tide, adult and larval *Aepophilus* shelter in rock crevices or in the holes drilled by rock-boring molluscs. They emerge at low tide to feed, searching for prey in seaweed. *Aepophilus* tends to be most common on the lowest part of the shore and therefore spends most of the tidal cycle under water. Consequently, it is more often seen by marine biologists than by entomologists.

Brian Nelson
Portadown, Armagh, UK

7 Sand, saws and the Klingons

❓ Written in the sand

Can anybody help us explain what caused the phenomenon in this photograph? It was taken on the Sands of Forvie nature reserve in north-east Scotland. The weather was dry, though there had been rain the day before. The photograph shows the lip and face of a small mobile sand dune. The slope of the dune face was approximately 45 degrees and the sand was dry at the time the photograph was taken. The image shows a section about 1 metre wide. The whole pattern extended about 20 metres.

Martyn and Margaret Gorman
Newburgh, Fife, UK

Moisture from the rain retained by the striated slip face of the dune would have given it a firmer consistency than the sand on the top, which would have dried out more quickly. What happened then would have been quite startling for anyone lucky enough to witness it.

The loose dry sand on the dune top would have started to move, pushed by gentle winds. As it spilled over the edge, the sand cut this series of parallel gullies, creating the herringbone pattern. The uniform spacing of the gullies is related to the even flow of the wind dynamics on the edge of the dune. Gullies would not have been able to form in the loose dry sand on the top of the dune.

At the dune crest, erosion would be occurring along a hemispherical front (as you can see from the curved tops of dunes in a desert) as loose sand fell into the gullies below. This sand would have flowed down to the bottom of the slope to form a fan or ramp at the foot of each gully (visible at the base of the photograph). Once started, the whole process would have taken only a few minutes. The gullies and ridges are very fragile and would soon be buried under sand as the dune face moved forward. The photographer was very lucky to see them.

I have seen this phenomenon on a grander scale in Arabian and African deserts where holes up to 6 metres deep have been dug in sandy soil. There, surface soil with some structure and a moderately hard consistency lies on top of poorly structured sand. The hard surface stops the face from collapsing but the whole system is inherently unstable and dangerous to observe from inside the pit.

A steady supply of drifting sand often blows over the rim of the pit, eroding the edge of the surface soil into intricate patterns, picking out minute differences in its constituents and etching out old roots and so on. The loose material produced by erosion feeds into a network of fragile gullies cut into the underlying sands. Where there are harder layers lower down, sand will flow down the first gully, cascade over the next hard layer as a 'waterfall', race down another gully and then broaden out to form a fan at the base of the slope. Such a flow can be maintained for minutes on end.

This type of flow produced the pattern that is seen in the photograph, and is beginning to be recognised as of considerable significance in the geological record, according to Brian Turner of the University of Durham, UK, who has observed similar particle flows in sand ramp deposits in Jordan. Allan Treiman of the Lunar and Planetary Institute in Houston, Texas, has argued that similar examples on Mars are the result of 'dry granular flows'.

Gullies observed recently on Mars were attributed to the action of water. I believe this is way off mark. Having seen the particle-flow process in action on many occasions, I believe the Mars gullies are far more likely to have been formed by dry granular flow closely related to that shown in the sand-dune photograph.

R. Neil Munro
Dirleton, East Lothian, UK

❓ Shapes in the sand

A friend spotted these unusual pointed sand forms on a beach in northern Lake Michigan recently. My first thought was that the shapes had resulted from rain or waves that had been breaking over the beach, but I'm still at a loss to explain how the shapes might have formed.

Michael Edelman
Huntington Woods, Michigan, US

There are a variety of possible answers here that have two things in common: a liquid capable of binding sand together and erosion. Readers disagree about what the liquid is and whether it comes from above or below ground – Ed.

The beach surface may look bare, but there is a thriving ecosystem in muddier layers beneath. Grazing marine worms burrow through it and then excrete waste onto the surface. Gut secretions and microbial films bind each miniature 'mud volcano' slightly better than the bare sand around. Gentle erosion does the rest.

Nik Kelly
Liverpool, UK

The picture shows a layer of wet compacted sand at the base of a series of sand pillars reminiscent of upside-down mushrooms. These shapes are formed as raindrops or spray from the lake fall onto a layer of dry sand supported by compacted wet sand underneath. As the drops hit the dry sand, they penetrate it and flow downwards, forming a vertical cylinder of moist sand.

When the liquid front of each droplet reaches the wet compacted sand underneath, the liquid begins to flow across the surface of the dry-wet sand interface, as the sand beneath is saturated. If the wind picks up at some point, it can blow away the surrounding loose, dry sand, exposing the mushroom shapes.

Similar upside-down mushroom shapes can be formed by placing drops of water on lactose powders and sieving away the unwetted powders.

Karen Hapgood
Seven Hills, New South Wales, Australia

The structures pictured appear to be what sedimentologists refer to as 'mud volcanoes', and are formed when a waterlogged layer of fine sand or silt is covered by a layer of coarser sand. The weight of the coarse sand increases fluid pressure in the watery layer underneath. Because the wet sand is fluid, it can escape upwards, making a series of muddy piles once it breaks the surface.

This can be seen in the photograph: some of the shapes are darker, and therefore wetter, than the surrounding sand and are superimposed on the wind-formed ripples, showing that they have been deposited on top at a later stage. Such structures are sometimes found preserved in rock formations.

Matt Carrol
Nottingham, UK

The shapes look to be urination concretions from an animal that crouches to pee. My cats make these in their litter box. The small central cone is the original point of urination and the larger cone-shaped area is the wetted patch of sand. The urine then dries, and the solids glue the wetted portion into a lasting monument as the wind blows the surrounding dry sand away from the area. These concretions also may resist rain and waves to a greater degree than loose sand.

Bill Jackson
Toronto, Ontario, Canada

I found very similar shapes when working in a field camp in the Tunisian Sahara. They were created when I had to go out in the night for a pee, and used an area of soft sand behind my tent. The urine, if one kept one's aim straight, appeared to vanish down a hole, but must have encountered harder, less porous layers below the surface and then spread out.

It must also have acted as a cementing material, because a few weeks later, following a severe sandstorm which blew away all the soft surface sand, these 'Mexican hats' remained very noticeable.

Bob Fryer
Comrie, Perthshire, UK

❓ Shifty sand

On a beach in Malaysia I saw small balls of sand in patterns (see below and opposite at top) around some holes. I couldn't see an animal inside any of the holes. So what makes these patterns, and how are they formed?

Romayne Gallagher
Vancouver, British Columbia, Canada

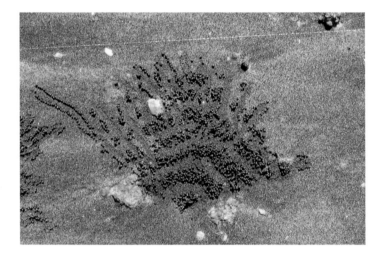

Clearly patterns of small balls on the beach are a common sight. So too are their creators. We've received reports from northern Queensland in Australia, Borneo, Goa in India, and Zanzibar. Here are two possible culprits – Ed.

The patterns are characteristic of the intertidal crab *Dotilla* (see photo opposite, at bottom). These very small crabs live in burrows, emerging at low tide to feed around their burrow entrance by scooping up sediment with their claws and passing it into their mouthparts. Organic material is filtered out to be

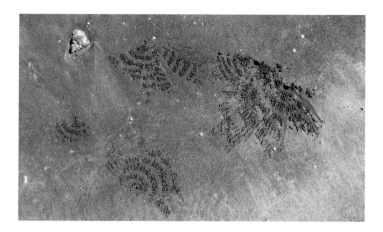

eaten, while rejected material builds up into a ball at the top of their mouthparts. Once this ball reaches a certain size, it is removed by the claws and passed underneath the crab's body. The crab then moves forward and repeats the process, dropping the sand balls as it goes and creating the pretty patterns seen.

Christopher Allen
School of Ocean and Earth Science
National Oceanography Centre
Southampton, UK

The photo shows a number of fiddler crab burrows. World-wide there are more than 90 species of fiddler crab belonging to the genus *Uca*. Generally they are small warm-water crabs that thrive in countries such as Malaysia, though they can also occur in temperate areas such as the Algarve coast in Portugal. Their common name derives from the male of the species, which waves a single enormously enlarged claw, rather like a violin-ist's bow, to attract females. Their other distinctive feature is their eyes, which are mounted on long stalks, allowing them to spot other crabs or potential predators from a distance.

Fiddler crabs forage for decomposing organic matter by scraping off the surface layers of the sand, which are rich in bac-teria, fungi, creeping diatoms, a variety of microscopic protists, plus dead organisms and faeces. The crabs ingest the valuable organic material, disposing of the unwanted sand by creating the numerous small balls that can be seen in the photo. Broadly speaking, the larger the crab, the larger the sand balls it pro-duces. In the photo it is evident that there were crabs of several sizes present.

Fiddler crab burrows are quite deep. This is necessary to allow access to water at any stage of the tidal cycle, even if they are some distance from the water's edge. When the sun is strong, and temperatures too high for comfort, the fiddlers retreat to the bottom of their burrows, emerging again once temperatures fall. They can spot people from many metres and respond to their presence by diving deep into their burrows, making it very unusual to see fiddler crabs at the burrow entrance.

From the photo one can see that fiddler crabs don't scrape sand in a 360-degree circle around the burrow. Instead they forage over 180 degrees or less. Fiddlers are highly territorial and need to keep their eyes on each other at all times; if they scraped a whole circle they would have to turn their back on a competitor. If foraging zones overlap, much time is wasted in displaying and fighting, so burrows are usually spaced out.

Each time the incoming tide covers the burrows the sand balls are normally washed away and the organic-rich surface layers of the sand renewed. However, some years ago in Penang, Malaysia, I saw a trail of forest ants marching out of the forest onto the sand and collecting the sand balls from the fiddler crab foraging ground, presumably as useful building material.

John Davenport
Professor of Zoology
University College Cork, Ireland

? See saw

I left a saw hanging up in my damp shed. Much later, I found it was covered in these seemingly random lines of rust. Why did they form like that?

Bill Adsett
Bristol, UK

The picture of the saw shows relatively large areas of bright metal. This suggests that the tool was originally coated with a protective lacquer, probably cellulose acetate butyrate. As lacquers age they lose plasticity and become brittle. The natural expansion and contraction of the blade with changes in temperature would split the coating, forming random fine cracks and exposing the metal surface to corrosion.

Allan Whatling
St Mawgan, Cornwall, UK

The tool pictured has undergone what is known as filiform (or underfilm) corrosion, a process which can affect metals that have been given a thin protective coating. The phenomenon has been the subject of extensive study since as long ago as the 1940s. Thread-like filaments of corrosion up to 0.5 millimetres wide grow in the coating at a constant rate of about 0.4 millimetres per day in random directions. They do not cross each other, and if one filament approaches another it will either stop dead or veer off at an angle.

The process occurs thanks to a differential aeration cell – which illustrates a neat little example of electrochemistry in action. In the head of each filament is a tiny pocket containing a concentrated solution of ferrous salts which can draw water from the atmosphere. Oxygen also diffuses through the coating and an electrochemical cell is set up. Iron oxidises under the head of the filament and a trail of rust forms behind the head.

Because of the nature of the reactions, the filament propagates with the head acting as a moving anode and the tail as a cathode. The potential at the head of a filament would be neutralised were it to encounter the tail of a different filament, so the only way it can propagate is to change direction.

Filiform corrosion can be minimised by applying a phosphate or chromate primer to the surface of the saw, although these are now being replaced by safer organic alternatives at an industrial level. But even the lowly shed owner can take proactive steps: wipe your tools down with an oily rag every now and again and you will stop the corrosion in its tracks.

I infer from the photo that the shed must have had a relative humidity of between 65 and 95 per cent. Any lower and the corrosion would not have happened; any higher and you would probably just see rust blisters instead.

Peter Barnes
Penarth, South Glamorgan, UK

⁇ Written in stone

I spotted these stones (in both photos) that had been used to make walls in Gozo, Malta. What aspect of composition and weathering led to their distinctive indented appearance?

Richard James
Cardiff, UK

The picture shows a type of stone known as a globigerina limestone, which is the traditional building stone of choice in Malta and Gozo. While easy to quarry and cut to shape, it has the disadvantage of being very prone to salt weathering, a process in which salts crystallise within pores in the stone and can eventually cause it to erode and crumble. In this case the salts are probably deposited by sea spray, although in towns they may be derived from pollution.

Weathering could be the result of the salts repeatedly dissolving and recrystallising. The effectiveness of this process is very sensitive to the size, shape and connectedness of the pores

within the stone, and the pattern of decay often reflects subtle variations in the grain size and porosity.

In a completely uniform stone, salt weathering often begins at seemingly random locations and excavates small hollows which, as they expand and meet, create a characteristic honeycomb structure. Where there are variations in the stone's texture, salt weathering will accentuate them.

This is what happens in many examples of globigerina limestone, which is a marine sediment made up of fragments of shells and other organic debris the size of grains of sand. Other organisms burrowed into the stone as it was laid down, and their burrows were subsequently filled in with debris that was slightly different from that of the surrounding sediment. These areas act as so-called 'trace fossils' in that they are a marker of biological activity. Their presence is picked out by salt weathering, when they distort what would otherwise be a regular honeycomb pattern of erosion.

Bernard Smith
School of Geography
Queen's University Belfast, UK

❓ Klingon technology

In a box of childhood rubbish, I discovered a rubber pencil eraser and a plastic disc – once fired out of a toy Klingon spaceship – that over the years had somehow welded themselves together. There was some transitionary rubber/plastic substance at the join. How did this happen?

Stephen Battersby
London, UK

I found details of the eraser on the manufacturer's internet site and it is composed of PVC not rubber. It is not obvious what the disc is made of, but it is probably a commodity polymer such as PVC or polyethylene. To make the eraser flexible it would be necessary to add a plasticiser – a low-molecular-weight substance which is dissolved in the polymer.

Over time the plasticiser can leach out to the surface. On contact with the disc it would have acted as a solvent, causing the two substances to stick together as parts of the polymer chains diffuse across the interface between them. This 'inter-diffusion' leads to entanglements, physically bonding the two materials together.

The level of penetration of the ends of the polymer chains depends on precisely which polymers are involved and the contact time between the two materials. A similar method is used to assemble plastic model kits: applying a suitable solvent to polystyrene surfaces slightly dissolves the polymer, after which the parts are pressed together, the solvent evaporates and a bond is formed.

In the case of the eraser, the solvent, in the form of the plasticiser, does not evaporate but instead migrates from the eraser into the disc, and this is what forms the transitory substance at the join.

Michael Nugent
Polymer Engineering Department
Athlone Institute of Technology
Westmeath, Ireland

❓ Thirsty jets

Sitting aboard a Boeing 747 stuck on the runway on a very wet day at Heathrow, I noticed that the engines appeared to be sucking up water from the tarmac in front of them. Strangely, the water rose vertically in a very narrow stream less than 10 centimetres wide from a point on the tarmac directly in front of each engine. Then, when the vertical columns of water reached a point about a metre in front of the centre of each engine, they changed direction to head horizontally into the middle of each turbine. They looked, in effect, like large walking sticks made of water (see photo opposite) pouring upwards into each engine. I can accept that jet engines suck in huge amounts of water on wet days but presumed they did it more generally from the air around them rather than somehow sucking it up vertically from the tarmac in such a specific way. What is going on?

Jennifer Gold
Madrid, Spain

The peculiar moisture cloud you saw is in the core of an intense vortex, much like those in dust devils or tornadoes. Mathematically, a vortex cannot terminate in free space, but must either form a loop, or attach to a surface – hence the vortex from the engine bends down and connects to the ground where it is held stably. These engine-intake vortices have occasionally resulted in accidents where people have been sucked into the engine. You can view a video clip showing exactly that at www.youtube.com/watch?v=YJf1okwb-5Y (in which the man luckily survives).

Ralph Lorenz
Columbia, Maryland, US

The air is being drawn evenly into the engine when you consider only the portion of atmosphere directly in front of the

turbine. However, the presence of the tarmac boundary means the bigger picture is rather different. Such a configuration can be thought of as a flat layer with the air being drawn towards a low-pressure region in the centre.

The conservation of angular momentum means that any rotation in the air mass intensifies as you get closer to the centre, just as it does with low-pressure weather systems. Regarding the case in hand, the speed of rotation near the centre became so intense and the pressure so low that water was lifted from the ground and along the vortex right into the intake. A tornado above a body of water can act in a similar way. The photograph shows the same effect, but here the extreme low pressure and high atmospheric humidity made the water condense out to form a visible mist.

A more familiar example for most of us would be to consider a body of water, such as a water trough, with a horizontal drainpipe removing the water from a short distance below the surface. If the drain pressure is low enough, the angular velocity of the water as it approaches the drain becomes such that a familiar air-filled vortex forms on the surface and extends down to a point just ahead of the drain before turning horizontally into it, just like the engine intake vortex, but with the fluids and airspace reversed.

Laurence Dickie
Brighton, East Sussex, UK

Acknowledgements

A special thank you to the production, subbing, art and picture research teams of *New Scientist*. In particular, thanks must go to Jeremy Webb, Kirstin Jennings, Paul Forty and Valentina Zanca (among many at Profile Books), Melanie Green, Judith Hurrell, Eleanor Harris, Nigel Hawtin, Ben Longstaff, Nick Heidfeld, Thomas O'Hare and Sally Manders.

Picture credits for copyright images: p. 3, John Hoyland; p. 15, B. G. Thomson/Science Photo Library; p. 20, Norbert Rosing/National Geographic Stock; p. 27, Bothus Lunatus/ John Downer Productions/Photolibrary.com; p. 30, Corbis; p. 36, Pierre Verney/Photolibrary.com; p. 60, Stephen Frink/Photolibrary.com; p. 61, Wang Jianwei/Corbis; p. 63, Tom Noddy; p. 64, Nigel Hawtin; p. 87, Motonobu Sato/Getty; p. 96, Mike Hollingshead/Science Faction/Corbis; p. 137, John Pontiel/ Photolibrary.com;p. 152, David Fleetham/Visuals Unlimited/ Corbis; p. 160, Thomas Volk; p. 180 (top), Image Quest Marine; p. 180 (bottom), David M. Dennis/Photolibrary.com; p. 199, Peter Thomas.

Index